Hand's End

Hand's End
Technology and the Limits of Nature

DAVID ROTHENBERG

University of California Press

BERKELEY   LOS ANGELES   LONDON

University of California Press
Berkeley and Los Angeles, California

University of California Press
London, England

Copyright © 1993 by The Regents of the University of California

Library of Congress Cataloging-in-Publication Data

Rothenberg, David, 1962–
    Hand's end: technology and the limits of nature / David
Rothenberg.
        p.   cm.
    Includes bibliographical references and index.
    ISBN 0–520–08054–8
    1. Technology—Philosophy. I. Title.
T14.R63   1993
601—dc20   92–39341
                                                                    CIP

Printed in the United States of America

1 2 3 4 5 6 7 8 9

The paper used in this publication meets the minimum requirements
of American National Standard for Information Sciences—Perma-
nence of Paper for Printed Library Materials, ANSI Z39.48–1984 ⊗

541 Nature diversifies
    and imitates.

41 Two infinities, the mean:
    Reading too fast or too gently you are unintelligible.

Artifice imitates
and diversifies.

Blaise Pascal, *Pensées*

The hand which scoops up water is the first vessel. The
fingers of both hands intertwined are the first basket. . . .
(One feels that hands live their own life and their own transformations.) It is not enough that this or that shape should
exist in the surrounding world. Before we could create it
ourselves, our hands and fingers had to enact it. . . . Words
and objects are accordingly the emanations and products of
a single unified experience: representation by means of the
hands.

Elias Canetti, *Crowds and Power*

Contents

Illustrations

The illustrations preceding each chapter are from *The Various and Ingenious Machines of Agostino Ramelli* (1588). Here are some of his own words describing each device (plate numbers are from the original work):

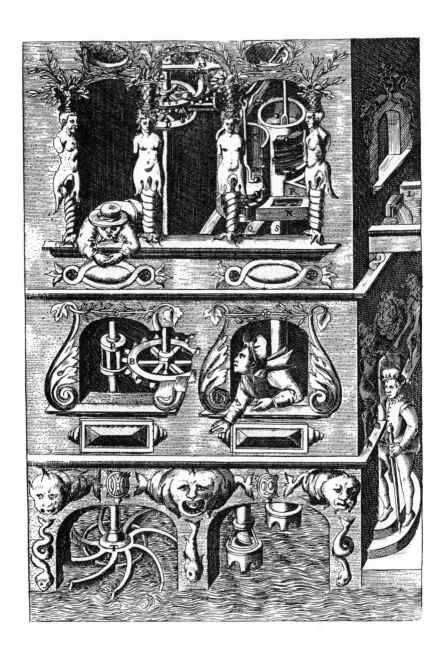

It begins with the hand—the grasp that pulls and directs; the movements enacted then fashioned out of material. Fingers trained to guide tools to reshape the world in our image, bridging the gap between those two infinities: human idea and tactile nature. Look at the preceding picture from *The Various and Ingenious Machines of Agostino Ramelli* (1588), a landmark work in the depiction of the inventive range of technological imagination. Gazing back several hundred years into the record of history, we can imagine what it might possibly do.

Water raised up from a river by means of the energy of its own current. Wheel T turns lantern P, turning gear B; wheel D and lantern C cause bellows K to draw water up through pipe S into channels N and R. A blueprint from the past, technical documentation for an almost practical invention. But there is more to the image than practice. See how the building is decorated with the fantastic: legless human bodies twirled on stems; gargoyles, goblins, fish and worms—ornamentation for the mundane and the mechanical. Machines have always been close to magic, and never far from emotion. Watch the three faces of those guiding its operation: they exhibit suspicion, anxiety, and reassurance. Natural flow is diverted through the implements of humanity. From intention to draw water away from its natural course comes the definition of the human course: bending and defining nature at one and the same time.

Technology is the totality of artifacts and methods humankind has created to shape our relations to the world that surrounds us, modifying it into something that can be used and manipulated to submit to our needs and desires. The world changes as we learn to see it in new ways. And the way we see the world depends on how we use it. We know no definition of the world outside our definite, practical engagement with it, yet the changing context continues to guide all attempts to remake it anew.

I will begin each of my chapters with a drawing from Ramelli's book and an analysis of the relevance of each image as a frame for the subject of the chapter. Or, each essay might be seen as an extension of the picture, discovering what an image comes to mean after four hundred years. I have chosen this one book to illustrate my reflections not only because the engravings are so rich in detail and ingenuity but because they reinforce another of my points: that technology has been essential to human definition of context for a very long time. There is no need to emphasize only the most contemporary of technologies in the search for this significance. We have the same problems Ramelli had in relation to the way technique transforms our identity. They are only more blatant today.

Still, technology was one of the first areas of human endeavor where the notion of progress came to light. Agostino Ramelli was a military engineer whose services were held in great esteem by the rulers of sixteenth-century Europe. His illustrated book of inventions became one of the most celebrated volumes of the period, and there is evidence that many of these machines were actually constructed and put to use.[1] Though the pages exist for posterity, Ramelli knew that the strength of his profession lay not in the durability of its creations but in the awesome expansion of possibilities each new machine might inspire:

> Indeed, to begin with Adam, the first father of humankind: every method and effort which he used to secure the earthly necessities by building stubble-covered huts and raising tight shelters to protect himself from the inclemency of the sky . . . — all this derived from the mechanical art. Unlike the winds that storm out of the cavernous centers where they arise and with their great fury rend mountains, open the earth, break thick walls, topple high towers, and dash wide-sailed boats into the vast sea, but then gradually weaken and lose their impetus and finally vanish; *on the contrary,* the mechanical art is like those great rivers that are small at their origin, but continually in-

crease as the many winding streams flow into them, and the farther they travel from their source, with so much greater breadth and fullness of water do they empty their heavy liquid burdens into the sea.[2]

It is not entropy that rules human progress but the continuing surge of innovation. Ramelli knew that the world was changing upon the transformation wrought by machines, and this was, for him, a direct contrast to the forces of nature, which likewise rise up, but also run down. Humankind, he thought, would soon triumph over these dissipative powers of the surrounding world. He glimpsed the change, but did he wonder what might guide it?

Tools change the range of humanity and the direction of human development, yet there must remain some goal outside of technology which machines should help us to reach. A common answer is mastery or control of the world, but recent events have shown that this kind of goal is not enough to prevent us from devastating the Earth in the attempt to take hold of its reins. There should be something else to strive for.

Here I will try to consider what "nature" would mean as this goal—not the irrevocable, essential forces of nature which make our evolution possible but a nature in which we struggle to fit, even *after* we have passed through the stage of exploitative civilization. This is not a question of looking back, but of reaching forward to imagine that we can find a home in the world after expending so much effort on its transformation.

I come to this project after several years working on the foundations of deep ecology, a philosophy of the relationship between human and natural which urges us to fit into an enveloping, moral sense of nature which is larger than any human purpose and which stands for the Earth or for life as a whole.[3] After a time, I began to be suspicious of this approach, as it tends to narrow the meaning of nature to cover only those facets of the world which may be seen as free from human influence. My intuition is that nature is far more important to humanity than this—a context which we discover when we touch, use, and change it. Hence, I turned to the study of technology, the way we delineate the world according to our purposes. We are always involved in reconceiving nature, whether we know so or not. This nature is more tangible than any imagined wilderness empty of the human gaze.

Technology does not exist without the human intent that drives it.

And "nature" is a meaningless term apart from our will to define it. A theory of technology will be outlined here as an extension of humanity, examining how the use of tools tries to realize human intentions, while transforming these intentions in the process. We end up with technologies that suggest new uses for themselves, thus evolving our original purposes. What we want to do is changed by what we can do—technology never simply does what we tell it to, but modifies our notions of what is possible and desirable.

Much contemporary criticism of technology tends to fall into one of two camps. The first fears technology as a monolithic megamechanism, rejecting it as a threat to our humane being. The second also imagines technology to be an entity separate from humanity, but one that will most fulfill human purpose when it is allowed to pursue its own inner logic. It will surpass us, and so succeed in dominating nature where human-driven action would naturally fail.

Each of these positions, however, imposes a generalization that blinds its adherents to other possible ways of looking at the problem. The idea of technology as a "supermachine" is simply too general to cover the whole slew of different and diverse kinds of instruments and implements used to shape a meaningful and comfortable world. We get little sense of how to deal with the implications of any individual machine if we simply make it part of a great, frightening conspiracy slipping out of our grasp, engulfing us, impossible to escape or turn off. A successful explanation of technology should not blur saxophones and motorcycles, nuclear power plants and ball point pens, all into one wrong turn in the story of our species.

The idea that technology will take over for us begins with admiration for machines that render human strength a luxury, but it forgets the romance of the struggle: steam hammers may dwarf John Henry, but how many songs are sung about them? Computers can beat all except a few grand masters in chess, but do they enjoy the game? Machine-built products are more consistent and exact than those hewn by hand, but does this make sewing machines better than knitting needles? Machines do not want to do anything, nor can they reflect on the wider significance of their activity. They work upon human and only human intention, and as such they serve to extend humanity outward, propelling us toward some uncertain goal whose appearance changes as techniques change. The question here is: how much is this end independent of the means used to locate it, and how much does it change as the range and security of our lives change?

From the first mortar and pestle, technology has been an expansion of human presence in the world. The lathe, the clock, the steam engine, and even the digital computer remain extensions of human means of manipulating the world into something we can keep track of and understand. No machine stands apart from its creator, no tool makes sense outside of its use. However, the device does assert a certain amount of independence when it begins to suggest other uses for itself. The packed dirt road allows us to travel further with horse and carriage, and later we build vehicles to travel more efficiently on paved roads. These in turn require better roadways, and soon it is easier to travel quickly to great distances from our homes. Where shall we go? The highways cannot tell us. Likewise, human ability to draw and paint images increases with the availability of chalk, pigments and appropriate implements. How do we learn what to draw? The paintbrush does not know.

The problem is that technological installation of human presence in the world invariably involves a series of choices. Choosing particular avenues of action necessarily closes others. As technical decisions are made, the original intention is fast channeled towards those possibilities which the technology admits. The more complex and encompassing the tool, the more it implies a unique, peculiar way of thought. This is why the construction of a simple rectilinear cottage leads to skyscrapers and why successful programming of a computer leads us to imagine that our own mind works the same way.

Clearly technique is not *merely* a means: just when we think the problem is solved, the machine reveals new troubles and possibilities alike. Technology wraps us up in its circle. This cycling path represents what it means to be human in all kinds of contact with the external world based on the dream of rational, planned order. Does any single external plan guide the wish? Theodor Adorno writes that techniques can "speak in a way which has nothing to do with the deliberate communication of a human message. . . . What looks like reification is actually *a groping for the latent language of things.*"[4] What Adorno is suggesting is that, while technology alienates the autonomy of the individual, it also seems to express yearning towards *nature itself,* some meaning inherent in things themselves beyond our own special use for them. This "latent language" would be a hidden logic of order in the world as a whole, beyond the pragmatics of human application. This is where technology not only does things but serves to explain them.

How is it that what began as tools of survival have come to offer us escape from the human-sided view of the world? How does technology seem to be a foothold to a place beyond our reach? Can our own creations still promise an absolute that is more than what we make, a world greater than our interest in it, a *cosmos* in itself?

This present work is then a *test* of the idea of nature as technology reveals it to us. Just when we think we have reached nature, it appears to have moved. Nevertheless, I hope that studying our practical engagement with the world may lead us better to understand what is in the world—as a truth that results from practice and art. If not absolute truth, then perhaps as much as we are meant to know; a human, living, and limiting answer reached through work, not just contemplation. The study of technology is not merely the study of its successes but also the study of its failures—one way we know nature's limits and our own.

Limits are what hold us back, but they also inspire humanity to keep striving to do and to know more. Nature is the most powerful of limits because we want both to be at one with it and to control it. In the race to expand our place within the laws of nature, we change what nature is so that it always remains beyond us, something to wish for, still far away, just out of reach.

While the computer represents a culmination of the reimagining of the world from within, another side of our technical prowess promises something much more immanently dangerous: destruction of our physical environment, either accidental or intentional. The possibility of global devastation through nuclear war or widespread global warming is real. Either would imply a negative judgment on the sum total of technologies: unplanned, unintended, yet dangerous without precedent. With our unrealized potential to destroy the planet, we reach the limits of nature long before we are able to comprehend them.

When the world becomes nothing but a thing to be destroyed, it loses its rich ambiguity. We react to the situation with recoil in fear of ourselves or assent to our inability to cope with the implications of technical prowess, and we give up, or give in. It is no longer possible to dream of immortality, as collective suicide appears more plausible or even inevitable. Does such an envisioned end cast a dark pall on the entire notion of technology, which now extends humanity only toward death? Belief in this vision affects us long before we learn whether it is correct. The world no longer appears without end. Any desire to wage war, now more than ever, needs to be tempered.

A sense of reverence for the Earth needs to check our technical progress, so that we do not simply wander down the path of ecological destruction. Yet we cannot continue if technology as a whole becomes the villain. We have always and will always need tools, in some form, to complete our existence. Some technologies enhance our place in nature, and these are the ones we should choose. We also need to be wary of how much technology closes off as it advances along one path to the exclusion of others. Yet so many of our actions change the very meaning of nature. This is the recurrent paradox of *techne*. Humanity is extended outward, but we do not reach out into the void. No—we methodically approach a context which will be ours only if we work to discover and maintain it. The fragile environment is easy to lose, even with at least one species upon it that has paused long enough to reflect on the course of change. Hence our need for a normative guideline of "appropriate" technology, where appropriateness is defined in terms of how well the whole of humanity can extend itself into the environment without destroying it or losing the ability to see beyond our own narrow interests.

Technology as the extension of humanity is the catchphrase for this entire work, and in the end one might wonder if the desire to salvage technology from its excesses does not stem from an arrogance that still places too much faith in the surge of technological progress. But the acceptance of technology need not imply denial of human ideals. Technique does not replace awe, and the universe remains more than what we make of it, or what images we supply to explain it. It will always be so.

I have tried to write a work of literature about technology, not only an essay in philosophy. Hence my choice of words is motivated not only by a wish for clarity in argument but by the flow of language as well. Sometimes these aims conflict, as the long historical feud between logic and style would suggest. But I do believe that philosophy need not be written in a manner too turgid for any but the most dedicated readers. If the subject is supposed to be important for many kinds of people, then it should be accessible to them as well.

Although this book claims to consider all kinds of technology, it does so from one very particular perspective, which certainly will not shed light on all mysteries of the machine. I am concerned with how tools are apprehended and used by *individuals* and with how these actors are changed by their involvement with various techniques. Larger issues of how societies or organizations control and are determined by technology are outside the scope of this work, but not because my approach is blind to them.

The logic of the work should show that, after classifying technologies into different types of extensions of human prowess, we discover how to guide an understanding of the limits of humanity's place in the world. Codifying our extensions helps us realize just how far they may be taken. If the system holds together at all, the reader should, by the end, be able to consider how any particular technology might bring our actions closer to the limits of the world and to determine what possible uses of that technology might lead to a violation of the integrity of the world. If these limits can be discovered in the properties of tools themselves, we will have a guide for living rightly on this planet which will help enable us to complete our purposes without destroying the world that makes them possible.

I am grateful to the Philosophy Department at Boston University for always finding room for my ideas, even when they seemed to violate the most basic of academic conventions in style of language and argument. Thanks to Robert C. Neville for finding something to say about every chapter to suggest that it represented *some* progress from the last. Thanks to Victor Kestenbaum for selling me the five dollar TV in chapter 3 and to Erazim Kohák for demonstrating how the reason he threw away his refrigerator up in New Hampshire proves the relevance of Plato's divided line to today's technological crisis.

I am also indebted to Arne Naess of the University of Oslo, who, in addition to finding me numerous mountain huts to think in over the years, always tried to warn me to stay out of philosophy until he announced one day that I was too deep within it ever to consider escaping. And to Roger Shattuck for urging me to stay out of the university altogether. (I'm one of those still on the edge of academia, continually testing the waters to see if it will permit freedom to explore without letting us get too lost or ineffectual.)

Thanks also to all those who have offered their helpful comments on the manuscript: Rabi Abraham, David Abram, John Canaday, Ed Dimendberg, Paul Durbin, Harold Glasser, Sigmund Kvaløy, David Landes, James Marquand, Jo Ann Miller, Justin Rosenberg, Dan Rothenberg, Ingrid Scheibler, Joyce Seltzer, Rex Welshon, Langdon Winner, and Michael Zimmerman. Thanks to Peter Kosenko for his superb and learned copy editing. Thanks to Bob Paslay of Symbiotics, Inc. for his generosity with information services. I am grateful to Tod Machover at the MIT Media Lab for inspiring me to conduct certain experiments that fueled this inquiry, and to my parents for never letting me forget the wilderness whenever caught up in machinery. And finally, endless gratitude to my wife Stephanie, whose presence in my life for the past five years has given me enough excitement and tranquility to embark on a project as far-flung as this without ever feeling bored or alone.

Cambridge, Massachusetts
September 1992

1   Unexpected Guile

You open the door in the gate and there it is: the machine outside the garden. What's it doing there? Pulling heavy weights along a track. Increasing the power of human leverage through the turning and hoisting of gears and pulleys. You are gently closing the door, and seem to be gazing away from the tool, somewhere off to the left. Forgetting the machinery, sweeping the eyes across the landscape. This unwieldy device might be an intrusion upon the rolling, tended hills. But it's been brought there, just beyond your door. You need that block pulled and lifted. And this is the technology to do it. It finds a place within your field of view. A slight smile—you need the machine to get the job done. Then your world is changed.

Technology. *Techne* + *Logos*: order imposed upon skill. When we organize our abilities, they become a structure which determines our world. We are accustomed to using the word *technology* in a total sense, symbolizing all that is mechanical in our society and thus opposed to a personal level of understanding: essential, constantly encountered, yet somehow out of reach. Our sense of technology could not have started out that way, and it cannot be penetrated if conceived as an impasse. No, technology is meaningful only as individuals use it, developing as it flows from the transformation of skill into idea. This first chapter traces the history of attempts to weave practical knowledge of how to construct and to make things with the ineffable sense of an overriding order. Technology stretches human presence further outward, making the world tangible by enveloping it in the web of our actions.

KNOWING THROUGH MAKING

People discovered early on that they could manipulate the world to fit their purposes. Philosophy wants to compare these purposes with the

1

order and purpose of the world as a whole. It is thus often opposed to technology: tools show a definite progress through history, while the same philosophical questions always seem to remain. Yet each tool brings with it the imposition of ideas. The merger of *techne* and *logos* is the logic of action, the order of art. It shows how our dreams are constrained by what we are able to do. The bond linking what we know to what we can accomplish is evident in some of the earliest recorded reflections on human place in the universe—the few surviving fragments of the writings of Heraclitus. The world, he suggests, does follow some elusive kind of order, but it is continually beyond our grasp:

> Of *this account which holds forever* men prove uncomprehending, both before hearing it and after they have heard it. For although all things come about in accordance with this account, they are like tyros as they try the words and the deeds which I expound as I divide up each thing according to its nature and say how it is. . . . For that reason you must follow *what is universal.* But although *the account is common,* most men live as though they had an understanding of their own.[1]

The italicized words[2] in this translation of portions of the first two fragments are attempts to render the elusive term *logos* into English. This Heraclitean *logos* is eternal, universal, common to all of us and everything around us, yet we are most often deaf to its presence. It's as if we live in perpetual slumber. To "wake up," we should learn to follow the order which guides the world, inclusive of our role in it. This is the *logos* of Heraclitus, the notion of order from which all Western attempts to claim systematic knowledge of anything are descended.

When we speak of techno-logy, are we asking for a systematic theory of practical action and artifice, or are we groping for a technique that is worthy of the order of the universe? The two are intertwined from the outset: practical knowledge and the ability to make give a certain sense of security and control that speculation will never warrant. A piece of the *logos* becomes tangible when it is tamed—when we work with it and are no longer terrified by it. When sure of a technique, we soon imagine what it implies about nature. Fire is a strange and terrible demon until we can light it and extinguish it. Once it becomes a tool, we wonder: perhaps this is what forms the universe? Or water, or air: these are not only sensed in the surrounding world, they are things which we *use.*

The wonder that engenders philosophy is *not* born out of times of rest, moments when we had leisure to spare in the struggle for survival.

No, it comes hand in hand with the human way of living in the world by changing it. It is thus difficult to separate our explanation of the world from our transformation of it. Heraclitus is wise enough to know that human language itself is strained and stretched when it is made to tackle such vast and vaporous concepts as the *logos*. But to bring this abstract omnipresence within our reach, he analogizes from the realm of familiar tools which do not scare us. What can he say to help us to contemplate the one and the many, the universal and the particular, the ambiguous and the definite? He needs images from the world of *techne*, of artifacts which we may literally touch, pull, and *grasp*. Here is why people are frustrated by the apparent indecisiveness of the world's rule:

> They do not comprehend how, in differing, it agrees with it-self—a backward-turning connection, like that of *a bow and a lyre*.[3]

What could this analogy mean? Both instruments—one of music, one of war—function through harnessing the potential energy within a drawn and taught string. So the universe seems built around a tension held and supported. We pull on the string firmly and gently, and either the arrow or the tone is released. These devices *work*: is it folly to suggest that the universe might work in a similar way? How might the bow and the lyre be *like* the order of the universe? Octavio Paz suggests that the lyre "consecrates man and thus gives him a place in the cosmos," while the bow "shoots him beyond himself."[4] For Lewis Mumford, the bow is an archetypal example of art and technics together, as it is both part of a musical instrument and a device that extends our aim by transforming the energy in the pull into the arrow's path.[5] Remember also Fragment 48: "The name of the bow is *bios*, its function death." In other words, the word *bios* can mean either "bow" or "life" in common usage; so the tool which kills is the equivalent of what lives. Once again opposites confront the truth through each other, in action and in speech. The question is an open one: we stop shooting, stop singing, and imagine a vast universe doing these same things, just like us.

This is but an early trace of a recurring kind of reasoning, as shall be shown in chapter 4. We see the progression of technical metaphor as the universe and its subset of inhabitants are successively seen through history to be *like* different types of machines—wheels, clocks, engines, arrays of digital on/off memory chips. The explanatory power of what we can construct should not be underestimated, though few have interpolated artifact into nature as subtly as Heraclitus. He has introduced the *logos* as an order which humanity ignores, but when he strives to make it accessible, he reveals its presence in the forces at work in the tools which we use all the time. The grand logic is explained through everyday things.

So from the outset, technology is essential to explanation of the world. Later, even such a skeptic as Socrates recognizes the exemplary value of technical knowledge: he mines it as the clearest vein of examples of how we progress from the particular towards the general—from specific skill to elusive truth. *Techne* is the clearest and most definite type of knowledge, because we immediately understand what it is for. "What result," asks Socrates, "does the art of the shipbuilders produce?" "Obviously," replies Euthyphro, "the making of ships." "Now tell me, best of friends, about the service of the gods. What result will this art serve to produce?"[6] With this second question, Socrates has analogized from the technical to the spiritual, forcing his accomplice to strain the term *art* towards the limit of what can be known. Art, the most common translation of *techne*, signifies the human ability to make and produce, and as *homo faber* we grasp its power at once. Knowledge which can create something useful to us may be justified by its end, and need not be questioned or mystified.

But it is not to be so easy—the virtue which Socrates so strives for will not stand to be encompassed by any learnable technique. No, the simple recognition of its presence comes only through ceaseless questioning of what can be accomplished, and doubt detaches one from the certainty of a sensible tool. A technique shows us how to make and control things, but it does not tell us why anything should be done. It soon becomes subsumed as a mere step on the road toward higher knowledge in the hierarchy of the philosopher: with Plato, technical knowledge is demoted in status, separated out from pure contemplation. Though he does recognize how good tools are internalized and accepted, he chooses to emphasize the gaping hole revealed when we admit that we cannot explain *why* we use them so readily without pause to reflect. If we cannot present any higher purpose for ourselves, why do anything? Plato instead looks for some certainty in detached, descriptive knowledge of the world, *episteme*, which might suggest an equally normative knowledge of the Good, the Virtuous, or the Beautiful.

Though technical knowledge might be an adequate analogy to suggest the kind of questions we need to ask, it gives no glimmer of the absolute to justify any hope. So Plato chooses to be inspired by geometry, a human discovery that seems to eclipse human limitations: a perfect circle is something we can never draw, but only approximate with our pens and compasses. Still, it exists as a tangible idea that transcends technique but may be easily grasped by the human mind. Quickly reason transcends prowess—our quest for the right way to live need no longer be limited by the approximate means we are able to master.

In his later work, the analogical power of technology returns, as Plato realizes that the idea of the technical end justifying the technical means is by no means as obvious as Socrates thought it was. The *Philebus* is a dialogue concerned with attempts to divide and conquer various types of knowledge. They are split up so that those based on random conjecture and guesswork may be weeded out from the exact, orderly, and "common" (in the Heraclitean sense) parts which could be universally called "true." Plato distinguishes the "element of numbering, measuring, and weighing" which constitutes the order inherent in any craft from that part based on "rule of thumb," enabling the practitioner to make the "lucky shots"[7] necessary to excel in the craft. Here, as usual with Plato's attempt at categorical knowledge, we are not exactly sure how he is relating the relative merits of each category to the single noble purpose of pursuit of the Good. Though the argument seems to demote once more this nonsystematic side of all actual techniques, there is a kind of latent respect for the *ability* to make those lucky strikes:

> Well, now, we find plenty of it, to take one instance, in music when it adjusts its concords not by measurement but by *lucky shots of a practiced finger*—in the whole of music, flute playing and lyre playing alike, for this latter hunts for the proper length of each string as it gives its note, making a shot for the note, and attaining a most unreliable result with a large element of uncertainty.[8]

Unreliable, yes, but still of the essence of art. It's not really so bad, as any musician will tell you how remarkable it is to be able to play passages quicker than one can imagine or plan them.

This section of the dialogue concludes with the stated aim that it has *not* been trying to discover "which art or which form of knowledge is superior to all others in respect of being the greatest or the best or the most serviceable, *but* which devotes its attention to precision, exactness, *and the fullest truth*."[9] Now does technology offer an order for machines, or new chances for surprise?

If we want to find order for the universe and human purpose within it, then we may be inspired by those aspects of technical knowledge which are organized, systematically and definitely arranged, possessing a *logos* of their own. Precision is recognized in the things we are able to master and direct before we can imagine it in the world outside us. The next question is whether or not we need to shun this penchant for "lucky strikes" which allows human improvisation to show through even the most codified of technical situations. As Plato's subtle way of embellishing distinctions suggests, one choice should never negate the other. He finds a place in his ideas and his rhetoric for both the comprehensible and the magical sides of technology, as stages on life's way, and as a profound analogy for a knotted duality in the rigging which ties all strands together, tangled between the random and the planned.

But such a wisplike, indefinite way of dividing and uniting ideas in disparate ways one after the other is frustrating to anyone who wants a quick-and-dirty answer. The students of Plato were aware of this, and they learned the value of taking a stand. Aristotle would never waft around the question, but state the simple and necessary truth of the matter: *techne* is a finite means, separated from the *logos*. Why and what it is used for is a separate matter. He takes his distinctions seriously, and when a concept is defined it is sharply delineated from related concepts: "*techne* is identical with a state of capacity to make, involving a true course of reasoning." [10] It is thus sharply distinguished from acting, which is governed by a separate faculty of practical wisdom, a virtue and not an art. This schism between the tool and the use may be seen as the root of so many problems engendered by technology. Most of us today, so influenced by Aristotle that we find many of his conclusions banal and obvious, tend implicitly to agree with his view of technology as a means to an end: it is the "true course of reasoning" necessary to keep machines and tools running.

By what criteria might this reason be called true? It is certainly not true in the sense in which the virtuous is true. Aristotle must be suggesting something akin to the precision which Plato seeks above, but in his characteristic manner he has sketched its edges more succinctly—the course is true if the technique works as expected, if it does what it is told to do, reliably and consistently, so that we may forget about it and concentrate on why we picked up the tool in the first place. Correct technology often works transparently as a means to realize the ends dictated by practical wisdom. More will be said about this in chapter 6.

But this invisibility is not to be carried too far: instruments do *not* accomplish their work without human guidance, like those oft-cited statues of Dædalus, which moved by themselves once completed, so perfectly did they resemble the living forms of which they were the image.[11] Yet they are an apt image for technology as well; *techne* imitates nature in its manner of operation, but completes what nature alone has only hinted at. Within the activity of art, all material is seen in terms of the purpose we are trying to realize, "and we use everything *as if* it was there for our sake."[12] This implies that the technical stance changes our sense of what the world can be *for us*. Attribute is perceived as resource; meaning develops out of the potential for use. According to Aristotle, since we too follow the laws of nature, what we make and build teaches us as much about the world as it extends ourselves:

> Now intelligent action is for the sake of an end; therefore the nature of things also is so. Thus if a house . . . had been a thing made by nature, it would have been made in the same way as it is now by art. . . . [which] *partly completes what nature cannot bring to a finish, and partly imitates her.* . . . It is plain that nature is a cause, a cause that operates for a purpose.[13]

Just what kind of cause is nature, within Aristotle's famous fourfold scheme? To the extent that nature is to mean the matter we may shape into our artifacts, it is the *material* cause. If nature includes the ideal schemata that make the building of things like houses possible, it could be conceived to be the *formal* cause. Nature as the prime mover is the original *efficient* cause for everything if neither God nor human free will are invoked. And if nature includes rules that determine how things grow, change, and reach their ends, it needs to encompass the *final* cause as well.

Aristotle's discernment of the varieties of cause is enough to encompass a wide range of colorations of the omnipresent concept "nature" in its most total form. But the imperfect analogy between the natural and the technical reveals an uneasiness always present in any attempt to spread nature too wide as an enveloping idea. Within technical process, the material cause still remains nature: we start with natural materials and refine them successively to suit our needs. Since we are the ones instigating a change in the order of things, realigning the world with our purposes, *we* are the efficient cause of any technology.

The two remaining categories are the ones open to the greatest range of interpretations. Presumably, if we imitate with our tools the manner of nature's operation, we apply existing natural formal causes to new materials, abstracting the laws of nature from their original context and realizing new potentials by applying them out of place. Need we say that those technical solutions which work are just those *permitted* by the laws of nature, included within its forms and essences? Under this conception, nature is the final arbiter of any technical stab at modification of our world. Or, if we abandon this view, we can name technology to be the *logos* of changing the world, specifically providing those forms which go against nature, perfecting it where alone it has been unable to succeed. It remains to be seen whether our attempts to alter the world are at all possible without some kind of *consent* from nature as the thing which precedes and guides all process.

Is nature a *sufficient* cause for technical innovation? Is nature strong enough to guide technology? Chapter 3 will discover that alignment with nature has always been part of theories that hope to direct technical development. Yet chapter 4 will show how nature itself changes as a result of different techniques that both extend human reach and offer new metaphors for the description of the surrounding, enveloping, present world and the forces which underlie it.

The final cause of a technical process can be considered natural only if human purpose is completely subsumed by nature. This would require either a broadening of the meaning of the word *nature* into a synonym for everything which could be created *or* a redirecting of the term so that it refers not only to what is, but to what will be, as we shape the world towards some goal which it was always meant to realize. Some of the dynamic sense of this process is beyond the scope of Aristotle's time, when explanation was more appropriate than prognosis for change. (It was not so dynamic a society as ours, in which tools seem to be evolving so much quicker than our capacity to make sense of them.) But he was still wont to justify nearly all his conclusions with the maxim that they are simply true "by nature." And this is a nature that includes morality as well as physical law, containing the ethics to guide technology as well as the materials and their relationships which make it possible.

In Aristotle, nature is an all-knowing concept that may be a synonym for the totality of the world as well as all the causal processes that operate within it. But *techne* represents the human penchant to go beyond nature, thereby completing our place in the natural world.

This concept contributes to the deemphasis on practical action within philosophy itself. *Episteme* takes over as the central human pursuit of pure wisdom: knowledge of what is unchangeable and fundamental, not what we can do to the world. Aristotle is important here not because he elegantly severs the concept of *techne* from the world it acts upon, but because he so subtly blurs humanity with nature when he hypothesizes how they work together: both follow almost identical principles, but automatically in the latter and voluntarily in the former. We choose what to do, while nature never worries. Yet the conflation of *techne* with *logos* imagines more order and design than Aristotle might care to admit. As we refine its compound meaning, we need to retain Plato's inquisitive stance while building on the deep roots of the metaphor of Heraclitus, which shows so poignantly how what is made may serve to explain what is thought to precede it. Unfolding a vision of technology as an extension of human essentials toward the limits of nature, these ancient historical views continue to cast a shadow on the great, unforeseen material changes wrought in our time and in the future.

STIR AND THRILL

Aristotle, as master of the art of distinction, tried to clear up those thickets where definitions tend to get lost in the tangle. "Experience seems pretty much like *episteme* and *techne*," he says in the *Metaphysics,* "but really science and art come to us *through* experience. . . . Now *techne* arises when from many notions gained by experience one universal judgment about a class of objects is produced."[14] Here technology is presented as some kind of generalization made in order to organize practical knowledge. It assesses what has happened to us, making disparate activity systematic. What is lacking from this conception is any room for the fact that technology changes what is possible to experience.

Is this a danger inevitable in any sharp definition of so broad a concept? If we instead consider technology as a part of all we do, we will see how it affects all aspects of our existence in the world. In this section, we examine two attempts to comprehend technology in a relational manner, as it streams outward from humanity, defining the world outside us, moving toward, transforming it. This dynamic view is appropriate for a world that changes drastically with each new day. Two thinkers of the present century catch this trend as it just began to be significant. John Dewey realizes that nature can never be severed from experience, which in turn suggests that the tool can never be considered apart from its application. Martin Heidegger points out that the objects in the world which we perceive as tools already contain within themselves signals of their role within human purpose.

When the world becomes a malleable entity which is no longer fixed nor absolute, the concept of nature enters a flexible phase, characteristic of a more dynamic philosophical world than the unified *logos* of Athenian dreams. Knowledge becomes something uncoverable by humanity as soon as subjective thought is made pragmatic. In the words of John Dewey,

> the assumption that nature in itself is all of the same kind, all distinct, explicit and evident, having no hidden possibilities, no novelties or obscurities, is possible only on the basis of a philosophy which at some point draws an *arbitrary* line between nature and experience.[15]

For what line can be placed between the two which would be more than folly? How can we know something about a natural world outside of the experiencing of the knowledge ourselves? This statement is obviously the product of a time in which humans have decided that what we know about the world is more accurately and more modestly what we know about ourselves in the world. Yet any virtuous humility is somewhat tainted by the power which the prerequisite of restraint gives us: if nature and our experience are one, then the world of our conceptions becomes simply what we can make of it. Does all mystery then retreat?

Not as long as nature-experience still includes things that may happen to us but are not explained. Not as long as we still do things that elude the grasp of analysis. "What is really 'in' experience extends much further than that which at any time is known," continues Dewey.[16] A philosophic vision that admittedly spreads outward from the human locates experience, not nature, as the primary totality. The conception of ends is, for Dewey, distinct from the Aristotelian notion of completing or perfecting nature. The contemporary world is as much a realm of technological objects—roads, houses, buildings, radios—as it is a natural world of trees, earthquakes, river courses and mud. We envision possible completions as we assess them all. We project towards our goals— "they are ends-in-view."[17] From our first glimpse of something useful, the qualities that inspire its application are as much within the thing as they are in us:

> *Things* are beautiful and ugly, lovely and hateful, dull and illuminated, attractive and repulsive. *Stir and thrill is as much theirs as are length, breadth, and thickness.*[18]

Here is an aesthetic vision of technical arousal. The world is not so much a given part of us as it is a happening that propels us outward. Building upon the primary field of experience does not mean we are stuck in the past; this is where technology comes into the picture.

A tool specifically anticipates the future, enacting primary relationships between external things. Because it acts directly from one external thing to another, the device reveals instrumentalities that connect the instances of the world. "The spear suggests the feast not directly but through the medium of . . . the game and the hunt, to which the sight of the weapon *transports imagination.*"[19] In Dewey we find the root of the idea that technical objects prefigure and vindicate their own usages within the moving sphere of human possibility. The virtue in this vision is the way that it surreptitiously does away with the dogmatic distinctions between the human and the natural, the made and the known, and replaces them with an exciting, swirling, *thrilling* continuum of activity through which the world is transformed by our presence and clarified through our techniques.

What's missing from this world of ideas in which all doors have been opened is any guidance in assessment. Sooner or later, another arbitrary line will need to be drawn—one between those technologies which realize the purpose inherently predicted by human experience and those which cloak our potentials by blinding our eyes to the truth no machine could show. Dewey is trying his best to objectify technology as an instrument of relations, to distance its significance from the anthropocentric notion of the tool as a solely human thing. His *techne* is decidedly not an extension of humanity, but a window onto the way the world connects to itself intrinsically, which we may partake of if we can mirror the play of natural forces with the certain understanding of our own tools and the principles that guide them.

In the course of discussing the way the human organism adapts to its surroundings, Dewey writes that the "secret of this technique lies in control of the *ways* in which the organism participates in the course of events."[20] It is not the participation itself which is altered but the choice of when to direct experience and when to let experience direct us. Though he recognizes the extent to which technology influences our ability to know, he does not venture to explain how the ways we exclude and include the surroundings exert profound changes on the nature of our intentions themselves.[21] Dewey is still too impressed with the clarity of technology, as were many of his contemporaries in the first part of this century. During those decades when engineering was the king of professions, it seemed that all disciplines might be improved through efficient and functional trimming down to essentials. Dewey is ready to begin engineering educational and community institutions; his admiration for the pragmatic remains too powerful for the pitfalls in this approach to come through. Still, his bold advocacy of sensitivity to the full range of qualities within things cannot be forgotten; what he calls the "Earthly environing world" is a rich and diverse vision of the environment we now need to argue for preserving.

Dewey's approach resonates with others that ask for the elucidation of the world's own attributes before any definition of the human is attempted. Consider the early writings of Martin Heidegger, particularly those passages of *Being and Time* which deal with the distinction between objects in our defining surroundings which are merely present to us (*vorhanden*) and those which are ready for us to use (*zuhanden*).[22] The latter are things made by and for people, while the former possess their own independent truths. Those falling into the second category are understood as *Zeug*, equipment, "gear," or the totality of tools necessary to complete a task. One cannot speak of *an* equipment, and thus the *Zeug* is the sum total of techniques implicit in all human-made things. We have built them, so we know what they are for.

His classic example is the hammer, which does not fulfill its purpose or meaning until it is hamme*ring*. The more nails we drive into the wall with it, the more we put it to use, the more familiar we become with its actions, the "more primordial does our relationship to it become."[23] Primordial in what sense? For Heidegger, immersion in the *Zeug* becomes increasingly intuitive, requiring less outright thought, bringing us nearer to our goals while we forget that the tool is in our hands. The world with which we deal most primarily will not be the world of implements and their specifications, but the work that needs to be done with the hammer. It is not the product, but the act of building, which matters most.

Zeug is part of the *Dasein* of human existence, a Being that encompasses the presence of all parts of the *Umweltnatur* that surrounds us. Technical entities that appear in this world possess a kind of Being that is immediately there for us, ready for use. But if one looks into the proximate production of these objects, one could equally say that they are manifestations of human attitudes to the world, made concrete as they are constructed, remaining so obviously *there* in the world *for* us because it is we who have construed their forms and then built them. Consider Heidegger's other examples in this light:

> In roads, streets, bridges, buildings, our concern discovers nature as having *some definite direction*. A covered railway platform

takes account of bad weather; an installation for public lighting takes account of the darkness. . . . When we look at the clock, we tacitly make use of the "sun's position." . . . When we make use of the clock-equipment, which is proximally and inconspicuously ready-to-hand, the *environing Nature* is ready-to-hand also.[24]

The disturbing tendency of these peculiar turns of phrase is to make the world-as-we-decide-it-is some absolute quality of the world as it presents itself. Bad weather is only bad weather *for us,* and darkness is only to be remedied *for us.* The aspects of the sun's position which we choose to measure are only those that help us pursue uniquely human concerns. Other organisms are sensitive to different degrees of the change within time, and they run according to their own internal clocks. Heidegger seems to confuse the issue here: is he talking about the way the world appears useful in itself, or only from the limited human perspective? (Can we ever have any perspective wider than the human one? That is the question underlying this work. Keep asking it.)

Keenly separating those aspects of our world which exist in themselves from those which reach their full being through our use of them, he still treats them all as objects in the field of perception, not sufficiently emphasizing that what is ready-to-hand is often what is made by us. He does not separate what humans do to the world from what it means to be human. This is because he is most concerned at this point with elucidating the meaning of human Being, which he calls *Dasein,* as a complex "field." Later, in chapter 3, it will be necessary to return to Heidegger as he clarifies what is especially transformative (and perhaps unnatural) about the whole of *modern* technology. Chapter 6 will discuss Heidegger's later suggestion that we release ourselves toward the world as an antidote to overemphasis on the machine. His viewpoint evolves as the century progresses, yet a strain of ambivalence remains.

Where Dewey tried meticulously to evoke the wonderful, promising continuum between natural events and human actions, Heidegger tries to uncover our enlightened position within existence by beginning with wonder at the plethora of objects within which we live. By deemphasizing the human element behind the drive to employ techniques, *neither* considers technology to be much more than facts in the midst of their respective explanations of the world. In the pragmatic sense, neither has a theory of how technology *works,* only how it fits into their respective total views.

Neither of these writers devotes enough attention to the ways in which specific technical practices change the world. Each is firm enough in the philosopher's privilege to believe that thoughts can and certainly do change things: "Every thinker," writes Dewey, "puts some portion of an apparently stable world in peril and no one can predict what will emerge in its place."[25] Such is the supposed power of ideas, but what of tools that extend our range and redefine possibility or even perception? How might a change brought about through technique differ from one which is solely the rearrangement of ideas to explain the world?

To make sense of this question requires a more careful consideration of the difference between what is understood and what is done. It suggests a distinction, however subtle, between what humanity is and what humanity is capable of. It is into these waters that we all will step: once, twice, and then many times.

THE CIRCLE OF INTENT AND RESULT

Technology seems to make us larger than life. Just as powerful is the basic idea that the world exists to be bent toward our purposes. This world is revealed to the extent that we can turn it toward our designs. Technology as a topic for philosophical speculation quickly veers toward the all-encompassing, emerging either as a total metaphor for our whole machine-guided civilization or as a neutral means in the mess of clouded, shaky goals and ends. Here I want to shy away from abstract, global pronouncements and to investigate how the tool specifically influences the range and direction of our thoughts and aspirations.

The more we learn about how to use an instrument, the less we think about it as we use it. It becomes like an extra limb, a new way to reach out and change the world. But what is it precisely that is extended? Not simply an internal human idea, but an idea to act, a thought that engages the world, making the possible actual. The more we understand of the tool, the more ways we conceive of how it may be put into practice. Our desires and intentions to act upon the world are themselves altered through the tools that we create to realize them. This is the essence of the philosophy of technology as human extension.

Joseph Weizenbaum describes it well in his book on the limits of the computer as an accessory to human reason:

> We must, in order to operate our instruments skillfully, internalize aspects of them in the form of kinesthetic and perceptual habits. In that sense at least, our instruments become literally

part of us and modify us, and thus alter the basis of our affective relationship to ourselves.[26]

Like our own arms, legs, eyes, and ears, instruments which we internalize can never be purely transparent extensions of any premeditated intention, since the successful application of tools modifies our desire to use them. Technology does not extend an essential human nature outward to successively inscribe an external nonhuman nature, but alters the meaning of humanity by turning our own intention increasingly towards those aspects outside us which may be shaped in greater accord with our ideas. We see the world as we make the world, and we make the world into what we have seen and imagined through the tangible construction of technical possibilities.

First, a stab at more precise definitions of terms which I have thrown about more or less as they are found in common usage.[27] I am not ready to lose the rich possibilities inherent in the ambiguities of "ordinary" language, but some specifics must now be stated.

Technology is the entire logic of artifice, the order of art. It has been demonstrated above to emerge as a historical concept once rules about the way we create and transform things can be identified. The *technique, tool, device, instrument, implement, technology* (in the singular form), or *machine*[28] will be the objects within the overall field of technology, those constructed things which enable humanity to make and shape the world in an ordered way.

What guides our action towards transformation of the world? First, the *desire* to do so. A human want, never easy to distinguish from a need. In its general form, emanating from us, outwardly directed, human *intention* focuses our attention toward the world as something to be made through our actions. This usage of the term *intention* is very different from that characteristic of phenomenology, as Husserl and Brentano speak primarily of intended *perception,* where human gaze is always directed to something, never abstract, always oriented to an object in the context of a universal horizon. The intention I am speaking of is more active, more akin to the common usage of *want.* It implies an attention to action. Any human act that makes a mark on any "outside" world also makes that world an extension of the human being who guides the change. In other words, the world that we make through technology is *humanity extended,* the hand's end.

What is meant by *extension* here? It is certainly not the metaphysical attribute of an entity which spreads out in space and time. Rather, the word is used to explain how intention operates in a dynamic, territory-expanding manner. It reaches out from the deciding being to leave a tangible record of the decision in the thing which is done. A part of human essence is evident in the things which we build, create, and design to make the Earth into *our* place. Techniques can extend all those human aspects for which we possess a mechanical understanding. Telescopes and microscopes can extend the acuity of our vision, because we know something about how our eyes perceive the world optically. But we cannot technically extend our sense of what is right, because we do not systematically understand how this judgment operates. And if we did, could we envision a machine to improve it? Only if morality could be conceived as a faculty, appendable by a device.

Technology as extension means that when we make something, we thrust our intentions upon the world. The result of an extended intention through technology is the *realization* of the intention. Sometimes this is something perceptible in the landscape, like a strip mine that has denuded a hillside upon the idea that the significance of the site is the coal which the trees and grasses kept hidden before. We intend the Earth to be a resource, and we drive our idea of resource into the ground itself. With technology, we turn the scenery into whatever we wish it to be: material for our sustenance, formerly invisible, now unassailable before us. What we saw in it it has become.

Other examples may be more subtle, as the realization of a technology turns back toward our intentions, showing us more of ourselves. We dream of a regular temporal pattern in the universe and observe its passage in the shadows and the stars. We conjure the clock as an image of time. It does not make or break any physical part of our surrounding world, but returns to us as a recurrent and regular reminder of a consistency that we hope is real, as discussed later in chapter 4. We come to rely on it to organize our lives, to fall into precise patterns from morning until night. The universe may possess light and dark, but without our penchant for meticulous division (see chapter 6), can we say there is such a thing as a nonhuman second, or even a millennium? Only *we* care about time, and extend it out to all things we believe to exist.

There is a variant of the concept of extension which serves to explain how a human idea spreads out to cover all we may perceive. *Expression* is the process by which a human idea is made into something else. Because it is a tangible manifestation of our thoughts, the tool and its outcome also exist apart from us—we can point to them while the idea remains unseen. The clock is an expression of our obsession with time. A skyscraper is an expression of our faith in gravity and our belief in the strength of rectilinear construction. The road is an expression of our desire to travel. These and all human-made things appear as expressions when we consider them in themselves, apart from the human need to cause them in some way. Perhaps the idea of expression is best understood aesthetically, as when we say that a particular poem or painting is an expression of sadness, loss, time, gravity, or the need to travel. When we assess the technique as an object, we examine what it expresses about its creators and their purposes. When we focus on these purposes and try to envision the technique as a means, we see it as an extension of what we intend to do. But it is never merely a means: loaded down with the baggage of our desires, the device turns back upon us and suggests new intentions, propelling its usage forward in time and towards more complex applications. This is what we like to call progress.

It is useful to speak of a technology as an extension of desires because this does not innocently detach the tool from its stated purpose. Following Dewey, they are bound together within the seamless web of experience. Why do we have telephones? Because we believe it is useful to be able to contact people all over the world by summoning their voices without leaving the house. It is an ancient dream, envisioned in countless myths in many cultures. Long ago there were drums that could beat out messages audible miles away. Then a technique is first developed to send messages by wire. All that the telegraph manages is a simple dash and a dot, the digital yes and no of the computer world. Soon voices too course through the wires, and finally they travel through the air without wires at all. Magic? Only if there is no systematic *logos* to explain it. With a global network of waves and satellites we may speak to anyone without stepping out the door.

But why speak? What is the intention which drives the device? Speech, as the basis for most human attempts at immediate communication, gains new powers as it is further disembodied. The telephone was initially a curiosity—who knew what to do with this sudden closeness? Everyone was like the child making her first phone call, not sure if the other end of the line was real. Now phoning is a fact of our lives. There is business by phone, love by phone, all kinds of feelings and thoughts coursing through the air at any given moment. Knowing we can reach almost all of the persons we care about by dialing a number radically alters the way we relate to people throughout the world. We speak of those whom we know not as friends, but as contacts; not as a community, but as a network.

A tool realizes a human inventor's intention, and the realization of this technique suggests new intentions. Those who use the tool begin with their own intentions, and the more they accept the technology, the more their desires are changed. The technique alters its user's grasp on the world. It is a circle of technological practice and technical innovation, a cybernetic relationship between predicted use and discovered use, a situation something like this:

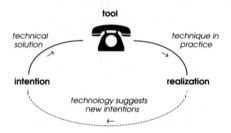

Figure 1. The circle of technology
in action

When we consent to apply a technique, we agree to let our desires themselves be manipulated and tested. One needs to do a lot of business on a telephone before conceiving the need for a fax machine. One needs to drive a lot and simultaneously to call many people before needing a car phone. And these gadgets need to be technically and economically feasible before they gain wide currency. Already their proliferation changes the way communication is conducted. The stranger thing is that we cannot now imagine life without them, even though they were impractical just several years ago.

What does the telephone express? As a working, designed object, the twofold nature of basic human communication coupled with a component to address anyone in the world by number. Microphone and receiver, one talks, the other listens. A society that did not value individual conversation so highly would not have any use for this ringing thing. Expressing the kind of communication we are most comfortable with, it transforms the extent of this need into the kind of talk it can handle: one to another, all aural, without body language or the effect of a mutually shared real space.[29]

Expression may seem here to blur with extension, but the two concepts need to be kept apart to remind us that technology is not equivalent to humanity, neither subservient to us nor prefacing us. Instruments never play just what we expect them to. Sophocles knew this, in the ode in *Antigone*, where humanity finds,

> in *techne*'s ingenuity,
> unexpected guile.[30]

The tool solves a problem, and then creates new and more thorny issues not dreamable before. Technology, unlike science, does not even claim to reveal larger truths about what exists, but hints at more ways for humanity to change the world. Born of simple need and want, it emerges as an agent of human evolution.

THE INTENTION MOVES

The image of the tool within the circle of intention reveals both instrument and inspiration. But this is not the whole story. Technology as a whole appears to *progress*, extending humanity, continually reconfiguring our abilities as well as our knowledge of things within and without us. When we speak of the *evolution* of technology, we imply that the change wrought by it is systematic and total, moving from one place to another. This section considers the parallel courses of the evolution of devices and the evolution of ideas, setting the stage for an extension of the ancient tendency to use what we can make as a mirror to reflect a glimpse of how the universe itself might be made. This rhetorical device we have noticed in Heraclitus and Plato stretches fully unto the machinations of our own time, tying expression to understanding in a knot that cannot be unraveled because it is constantly being created.

Plato looked to geometry as the supreme example of an abstract kind of certainty which exemplifies the higher levels of truth which the human mind is permitted to catch. The perfection of the idea of a circle clearly outshadows the smoothness or roundness of any circle we might actually construct, supposedly proving the superiority of the mind to the hand. Certainty like the circle's inspired Plato to hope that all aspects of our mundane practical life might have equally crystalline guiding ideals. But how can we find them?

Progress in the search requires an attempt to examine the connection between the practical and the ideal, rather than simply to categorize them as part of a rigid hierarchy, with one superior to the other. The question "how did we *first learn* geometry?" was reintroduced into philosophy by Edmund Husserl over two thousand years later. He was intrigued by the observation that pure shapes seemed to be visible and identifiable in the surrounding world which we constitute as we live it, the *Lebenswelt* or lifeworld. Perfect geometrical shapes are not ideas that dwarf actualities, but inevitable consequences of our creative perception of this milieu. How do we move from imperfect image to perfect idea? How do we learn to extend what we see? For Husserl, the human technique of measuring is a *practical* discipline that gradually assesses sense impression according to a set of "limit-shapes" discovered through the directing tendencies of actual perception. In this view, the Platonic push towards the ideal is itself a *techne*, making as much as it is learning:

> Of course there is a limit to what can be done by the normal technical capacity of perfecting, e.g., the capacity to make the straight straighter and the flat flatter. But technology progresses along with humankind, and so does the interest in what is technically more refined; and the ideal of perfection is pushed further and further. Hence we always have an open horizon of *conceivable* improvements to be further pursued.[31]

Ideal shapes are nothing but the consequence of the practice of human *techne*: the ability to improve and the desire to place a numerical value upon a length or an angle. This inspires movement toward perfection, and the horizon of this motion is the perfect idea. It is never reached, but always present as the overall context. The more that is discovered, the clearer this context appears in our field of view.

The method of measuring itself may be called a technology if it is seen as a way to channel our thoughts and actions along systematic principles of order. The order may ask for perfect accuracy, but it should not forget that the perfection did *not* precede the technique. There was no absolute standard in place when people began marking the world according to "feet." All technologies make their results, so it is easy to lose the grip on the intention with which we began. "To the essence of *all* method," writes Husserl, "belongs the tendency to superficialize itself in accord with technization."[32] This *Technisierung*—"technifying" or "technizing"—refers to the process of how an intention becomes a technique, fleshing out the line in our picture between intent and tool. There is more to the world of our experience than method, and the fluid beginnings of an experience always contain aspects that elude the attempt to transform them into explanation through use. Husserl's theme is the place where we all always begin, not the context of perfection, but the infinite richness before interpretation: "The free, *imaginative* variation of this world results only in possible empirically intuitable shapes and not exact shapes."[33] Exactitude is a technique, not a quality, of experience. If we are to name it as an attribute of truth, we tie truth inexorably to what we can make and do.

Now Husserl was most concerned here with the elucidation of the originary qualities of this pregiven world from which the horizon of perfection may be deciphered with a code of techniques. Here I will not focus on the search for the universal pregiven environment of all humans. No, too many layers of technology need to be stripped away to imagine that. The issue here is not what life is like before technology, but how life becomes itself through technology—how a change in tools mirrors the refinement of the idea of human life.

To translate *techne* as "art" or "making" is not enough to get a sense of its progressive, dynamic effect on all of us and our world. In a world successively generated and regenerated by technology, the essence of change must be sought in the activity of *invention,* where new techniques are thought of, constructed, and put into practice realizing their prerequisite intentions, effecting new intentions, and changing humanity in this cycle. Invention has been central for many historians and philosophers of technology, because the development of new implements tests human creativity against the restrictions of an external nature. By selecting and extracting certain possibilities out of nature's potential, we compare our own plans with those permitted by natural constraints. If we succeed, we feel that we have overcome these laws, outpacing nature and liberating ourselves from some of its reins. Is this an accurate conception of what happens? Do we glimpse the horizon of perfection in invention, as Husserl would have it, or do we find a piece of the inner world of possibility?[34] If technological foresight becomes imaginative enough, its ends become more remote from the immediate incomplete patterns with which it began. Techniques solve problems and transform our desires, but the thought leading up to their development may also inspire intentions realizable only in a distant and dimly forecastable future. Technology evolves when it sees beyond immediate and codified wishes well before they may be enacted. It is through the imagination that technology pushes the farthest forward; it is in dreams of the future that the context for progress is determined.

But how teleological is this striving toward the currently impossible? Is it determined by a horizon of perfection, or a more adaptive process whose ends are difficult to pick out? We seem committed to the notion of progress without knowing exactly what it implies. Chapters 3 and 4 both consider how the history of technology proves again and again to be the one area of human endeavor where forward motion is easiest to observe—improvement in morality or sense of place is much less apparent through time. It is easy enough to spot the better machine, but the connection between this and greater human ends may be less clear.

Recognizing an evolutionary nature in the change of technology is not the same as concluding that technology is progressing through successive improvements toward some distant goal. Remember that the application of the idea of evolution here is a metaphor borrowed from biology. Testing the analogy more closely, note that the process of natural selection inspires change without providing a goal for this change. The life of a species in its environment may be conceived as a continual attempt to solve the same problem: how to mutually define a piece of the diversity which is subject and context. Neither organism nor environment exists apart from each other, and neither strives to exhaust the other, since each reciprocally guarantees the other's place. The same may be said for technical problems and the devices that solve them: situations and solutions change as they influence each other, but every new tool is not so much an improvement of earlier models as an adaptation to the new situation altered by the application of previous attempts. The mechanism for change, which may seem to us as we look back to be a beacon of perpetual improvement, does not constantly take a stand on the ultimate purposes it may effect; it affects the future through the moment with a subtlety hard to catch.

The wheel is surely among the most significant inventions in the history of humankind. It was first discovered or invented in Mesopotamia about the fourth millennium B.C. and rapidly spread East and West thereafter. Its earliest uses were ritualistic and ceremonial, then later military, but only on solid ground. The most familiar use of wheels as an essential component of vehicles designed for transport comes to fruition where a system of relatively smooth and extensive roadways may be put into place. In parts of Europe this proved feasible, but in the cradle of its origin in the arid lands of the Near East, the wheel was abandoned in favor of a more efficient and regionally appropriate means of moving people and supplies—the camel. In the wild shifting desert, where footprints and tracks last only as long as the calm between storms, rutted roads are impossible to sustain. And for nomadic groups always seeking a new place to move to, the predictability of travel routes was not a desirable innovation.

If a society abandons an innovation such as the wheel, does this mean that it is turning against progress? No, it means that theirs is a culture not enhanced by this particular technology. Wheels were also known to the Aztecs, but only on miniature figurines of clay animals, fitted with axles and disks to make them mobile. We do not know the purpose of these objects, but we know their builders understood the principle of the wheel. In their dense jungles, difficult to penetrate, it was not a device conceivable as the solution to a practical problem. It could not fulfill any imagined desire to move along roads. The intention to do so was lacking.

So inventions do not include within them any obligation to apply them to change the way we think and live. Radical transformation inspired by a technology must come from an unfulfilled pattern in the interaction between the technique and its context, which is made visible by the conception of something new. Such a notion of "technological selection" is similar to natural selection in that it suggests a model for great change in technology, as the many are weeded out in deference to the best. The theory of natural selection is an attempt to explain how life forms may change and transform themselves and their contexts *without any guiding force* behind this change. The one force required by the theory is the penchant for random mutations: selection proves a way for the system as a whole to deal with mutations. Historian George Basalla notes that in 1867 over five hundred different kinds of hammers were produced in Birmingham, England, alone.[35] Such technological diversity casts a new shadow on so many philosophers' attempts to use the hammer and the act of hammering as a single, unambiguous example of how a person uses and inhabits a tool. The fact that we have much less variety in our choice of hammers today suggests that only a few of these diverse designs proved to be adaptive in the long run, as machinery developed to take over many of the tasks previously performed by hammering.

The history of invention, though certainly wrought with countless "random" new ideas that seemed to come out of nowhere, is still the history of conscious human creativity acting in response to perceived problems of human life in cooperation with and distinction from the world around us. Behind any new human idea is a person or group, using imagination to conceive of a new way of realizing human purpose.

In setting the stage for a possible solution, the inventor may make use of all kinds of knowledge of different kinds of technology and science even far removed from the immediate predecessors to his solution. Unlike biological evolution, the new species need not unfold out of a slight or sudden mutation of the old. The domesticated camel is not a new form of wheel but an entirely new solution to the problem of how to travel from one place to another. It is also an answer that supports the need to go from one place to many places, by many routes, without having to retrace the same steps or alight at the same oases. The wheel, on the other hand, might drag us down into already trespassed ruts of treaded sand.

So the cycle of new and "improved" technologies need not be seen as a linear rise towards progress in the easier realization of human ends. It is closer to Ramelli's ever widening stream, the more tributaries do flow into the surge. Yet where will the course take us? A detailed examination of the way tools have been refined and reapplied through history reveals "*no* causal connection between advances in technology and the overall betterment of the human race."[36] But we are certainly a different people when more and more of our actions are mediated by machines. It is part the inventor's responsibility to realize that technology's effects are often independent of the reasons that demand them. Chapter 5 will show why. One should not shirk the responsibility of technological evaluation. Though we may give up the idea that a culture completely immersed in many-layered techniques is any better than one that works more directly with the organic world, we should not forget that our overall goal is to determine how to make the best use of technologies upon the awareness that they change us as we change them. As we learn new ways of applying our measure of systematic change through the work of our imagination, we need to retain a sense of just what part of our humanity is extended and what part is transformed.

RELOCATING INFINITY?

If technology affects the idea and purpose of humanity, human meaning is a moving thing. Technology changes something about our essence as it changes itself. This does not mean it is necessary to speak about human life before technology and then after technology, or to force the equivalent by picking some arbitrary line where technology influences enough to have at last a fundamental impact. On the contrary, we find it as a factor in all conceptions of humanity reaching outward in and to the world. We place ourselves in the universe based on what we can *do*. According to Goethe, the deed was first. Only later do we ask if it was worth it. Do our deeds transform ultimate purpose enough to alter our basic conception of our place, or just the *scale* of our confinement?

Remember the famous proclamation of Pascal, who tries to site a location for humanity based on a mean between the infinite and the infinitesimal:

> What is a person face to face with infinity? What is humanity in Nature? A *cypher* compared with the Infinite, [but] an *All* compared with Nothing, a *mean* between zero and totality. . . . The end of things and their principles are unattainably hidden from us in impenetrable secrecy.[37]

We are bound to the universe in a precise way by virtue of our size and reach, which give us a particular *portée* in the field of possibility, a scope that places some final limits on the extent of what we may know and do. We learn just what the world discloses to us; ultimates beyond this intrinsic link are beyond our power to comprehend.

Now does technology let us break these barriers? It defines our connection to the world from the beginning. As Paul Levinson puts it, "without technology all the world is either material and nonhuman, or human and unmaterialized."[38] Making projects thinking towards the universe, blending mind with matter. It is a precursor to any definition of reach at all.

So do more powerful tools remap our region, or do they only push us closer and closer to the predetermined limit, which Pascal called as inevitable long ago? Technology breaks the dualistic spell that conceptually separates thought from material. Ideas are made durable when shaped into implements, and our cognitive, tactile, and motile faculties are extended. How far? Will we grow and grow, unnoticed by the universe? It is tempting once again to check our advance with a new, still enigmatic notion of the mean. This mean is not so easy to spot: we are no longer sure whether what we see is us or not us. Or, can we perceive anything more than our techniques may allow us? Who sets the limit on whom here? The dictates of the universe or the acumen of humanity? Levinson, like most cautious commentators, does not take a final stand:

> The act of extension, then, is an act of transformation, in which aspects of the universe bordering on the infinite and the infinitesimal are transformed into human proportion. Or perhaps the

transformation works the other way, perhaps it is we who . . . are extended into the dimensions of the universe.[39]

So it is possible to adhere to an understanding of technology as extension without taking sides on who or what is extended. The telescope brings a glimpse of the distant starry heavens within our grasp, even while revealing places we could never physically reach if only because they are light years in the past. The microscope, on the other side of the scale, shows us worlds too tiny to be sensed directly. When we realize them, we understand that we participate in this level of life all the time, though necessarily beyond the reach of our consciousness. When it comes time to act at this level, as in microscopic experiments and operations, we function through a complex web of perceptual extensions that *reduce* the scale of our gestures by shrinking the effect of our intentions through machinery.

Obviously the wonder of this has enhanced our capacity to work on the world, but it has not reduced our awe in the face of the world—the more levels of infinity in view, the more we refocus our gaze. With technology as an essential mediator between humanity and Pascal's two infinities, we may perceive them to merge into one, standing at both ends as a circular limit to human aspiration. Pascal remains neither right nor wrong, while specialized instruments reveal how each infinity resembles the other. (Remember the story of the Incredible Shrinking Man, who found himself at one with the stars as he got smaller and smaller, dissolving at last in the scale of atomic orbits.) So the universe may be one at either end of itself from our vantage. It is no coincidence that the pantheist Spinoza was a grinder of lenses, as will be recalled in chapter 3.

It is not a tragedy that the ultimate end of things is hidden from us in perpetuity. It is a consequence of the fact that we know such things only to the extent that they are related to us through tools which reach from us toward them. Instrumental meaning opens successive doors. The specific parameters of our range are what changes as technique evolves, while the *fact* of the limit does not. "There is a limit!"—that was Pascal's cry. If technology is to do more than sharpen our conception of the world, it needs to harden the facticity of this limit, revealing it with diamond-sharp clarity. At the end of the modern world, we will see the human edge more clearly than ever.

Here we are, poised between antipodes of the conceptual Earth. Now it is necessary to list the ways we may move from our balanced foothold, climbing somewhere where we hope the vista will be explicit and vast. As the ways technology may extend us are enumerated, remember the nagging questions: "What do we grasp of this limit? How will we know enough of it to be guided?" *The machine is always there beyond the edge of our thoughts, in the garden of action, just beyond the walls of home. When it takes us to the top of the mountain, we are no longer sure we could have walked there ourselves.*

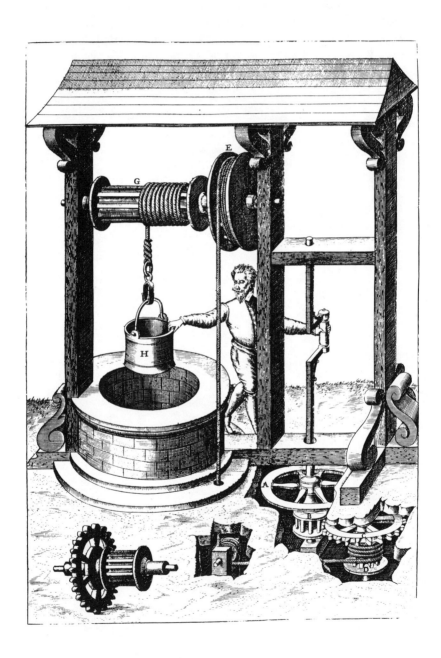

Hands are delicately poised on each end of the machine. The crank circling subterranean gears turned by the left, the right balancing the bucket as it descends off the winch into the well. Here are tools balanced gently by the grip, extending a turning motion to pull water deep out of a hole in the Earth. The device is driven by a man, but works only in a world already constructed upon our desires. The necessity of water comes deep out of the ground, a ground in which enough machinery must already be in place. Part of this machine extends the touch of the hand, part of it is only driven, with a motion all its own. The structure itself is the fragment of a built world, set in place by human design.

Dividing the tool into parts brings out the diverse sides of a technology, testing how much of it can be conceived as the extension of a human movement or idea. That is the subject of this chapter: the development of a system to classify technologies in order to see how far the notion of human extension can go.

The problem with any scheme that places knowledge into neat little compartments is that once we accept it we tend to ignore any experience that does not fall within its borders. And no classification system will explain everything.[1] Those who think technology encompasses everything may be disappointed. And if you prefer to narrow technology to the contemporary or the unwanted cutting-edge innovation, you will find the scope of the discussion is here much wider. Technology has a part in all that is made or done to the physical world by humanity, whenever we take up external material and call it our own by transforming it into something organized to further our aims.

Technology extends humanity in at least two ways. 1) It increases the boundaries of our world by enlarging our capacities. The sum-total of *techne* evolves to the extent that more and more of the world comes within human grasp: the universe comes nearer as the grasp itself gains in prowess. More and greater *action* becomes possible. 2) By extending our faculties through technical improvements, we are able to amass and process greater quantities of knowledge with greater efficiency. The human gaze becomes more acute. More precise *conception* becomes possible. Through the first course of evolution, technology allows us to do more, thus extending our presence on Earth. The second course allows us to see and to know more. The question here is to determine how action sustains thought and to decide if what we know can be separated at all from what we can do.

Physical action is clearly an area where technology extends our intentions. There is only so much we can do with the limbs we have. There are three ways of doing more. 1) Anything that builds upon those limbs meant to grasp, pull, touch, and strike is a direct extension of the limits of our bodies. 2) Then there are physical extensions of our desires which function independently from our own limbs but require us physically to drive them. 3) Finally, there are physical extensions that operate independently of our physical actions, embodying human alteration of the environment. These are those mechanized systems which, if simply set into motion, do the job. They are the technologies that seem least like extensions, and they are the ones most likely to conceal their human roots.

The extensions of thought are the tools which radically change the way we think and organize, without implying immediate physical effects. Once again there are three kinds. 1) There are devices that extend the scope of our perception. The faculties of sight and hearing are clearly extendable, with touch, smell, and taste more fixed. 2) Conception is enhanced, if not created, by the codification of abstract forms of reasoning, such as communicative language and mathematics, which may be considered technologies to the extent that they are constructed instruments that help realize our aims. 3) Memory may also be extended if we have devices that hold information *without* abstraction, in images retained to be sensed by sight and sound.

Actions do not lack thoughts, nor do thoughts shun actions. The division is useful just to help identify the range of tendencies that different tools exhibit, to determine how much can be included within the general designation "technology," and to test the value of the cyclical model. When technology becomes an ineffable organizing force on our lives, mental and physical, the boundaries between types begin to break down. If technology enables us to *deal* with greater amounts of information as it permits them to us, this would be evidence of the circular effect on intention. There may also be some human faculties which are *not* enhanced by artifice and machinery. Yet these are seldom independent from the transformative effects of technology.

Now through each of them one by one.

EXTENSIONS OF ACTION

Strengthening the Grasp

The simplest of techniques let the limbs do what they can begin but cannot complete. Direct extensions of the motions of our bodies expand the limits of our physical scale in the wake of the universe. We want to break the boulder into pieces with the sweep of our arms, but the fist is not strong enough. The unceremonious hammer takes the existing motion of the arms, turning potential energy into kinetic energy, smashing the rock into tiny pieces. The arm of steel saves us. Our hands would shatter into pieces if we smashed them against the rock with the same force which the hardened implement is prepared to accept. But the tool still works through the direct motion of the body, not independently of it. Aware that the hammer works, we are strengthened by it. Possessing it we feel more powerful than without.

The lever allows us to lift much more than our own weight, coupling mass with distance on the fulcrum. The spade allows us to dig deeper into the ground with greater resistance than the cupped hand. These inventions increase the extent of our strength, while others focus our movements. The knife enables the rough but determined movements of the hands to cut through all kinds of materials. The scythe makes the rhythm of turning into a swath that slowly cuts the grass. These direct extensions increase our confidence as we step, strike, and dig our way through the world.

What We Set in Motion

How far can we extend immediate intent? There are many ways we can set and control things in motion without directly extending our movements. As long as these devices still function under our direct control, they extend our bodily presence, if not specifically the actions of our arms, legs, hands, and feet.

The wheel fits into this category. No part of our body rolls naturally, but we find that it is highly efficient to move things from place to place upon rolling parts. We set the wheel into motion, by pushing it, by summoning horses to pull it, by attaching it to an "interface" (such as the pedals on a bicycle) which responds to direct bodily motion, or by constructing an independent engine that runs only when a person turns it on and guides it, as in an automobile. All these machines require close human control to operate, but this control is first the planning of the journey, then keeping the course with small indicating movements that guide the motion forward.

The techniques in this class operate with greater independence from human action than those in the first category. All transportation machines fall within this group, except any that could function purely on automatic pilot. Even the wheel's successor in the desert, the camel, fits in here as long as it is directly saddled by a human rider. But the minute a technology fulfills its function independent of a human impulse, it becomes more a fixture of our environment than a tool that needs to be handled to be effective.

A Separate Reflection

When a technique may be said to "work" even though no one guides it, it has become a physical embodiment of a human intent. It shapes the environment in our direction by extending a human concept into the workings of the immediate surrounding world. There are technologies which we only have to construct and maintain, not incite or activate, because they operate as a consequence of our bending of natural processes to achieve our way of life.

Once we have charted and dug out the irrigation ditch, the water will gently flow into it. Once the waterwheel is in place, the grain is ground up or hydropower is generated. Once the roof is placed overhead, there is no need to worry about rain in the night. We make the world into what we need from the world. These creations then act in themselves to make us more comfortable and less afraid.

This kind of technology is a separate embodiment of our extended wants. Because we can construct tangible manifestations of our ideas in a world of increased hospitality, we adapt the environment to our wants as much as we adapt to its constraints. Because technology can function apart from any watchful guiding eye, the changes it brings recede into the background of our lives. We take them for granted as part of the given lifeworld, which is never without generations of technical enhancement to ease our passage through it.

Several years ago Kali Peary, a Greenland Eskimo, the son of Admiral Peary and an Inuit woman, traveled to the United States to see the homeland of his father for the first time, at an age of more than eighty years. What impressed him most was the seemingly endless network of roads. "Perhaps," he said, "they have been here since the beginning of time."[2] Most of us do not think the roads we drive and walk upon have been here forever. But we do not remake them every time we use them. We do not need to assert control over them each time we travel. They are *there* for us, a solid image of our desire to travel farther and faster with less trouble. With roads from place to place, linking the locations we want to visit, the world literally becomes within our reach. The road extends our area of habitation and influence. It functions simply by being in place when needed.

This is how the Eskimo is correct, and why he was surprised: he may never have seen such tools that do not need to be set into operation to be effective. What he sees in the roads is a part of our constructed environing world. With roads we have extended a desire to communicate across the landscape, building onto it an image of this idea. A moose might come out of the forest, puzzled by the sight of the road, either cross it perplexedly or turn back the way it came. Its intentions are not realized by this strange new pathway. The more the Earth is moved in our direction through our extensions, the more it becomes an exclusively human world.

It is more difficult to call this independent sort of technology an extension in comparison with the two earlier types. But if our restless need for movement is not extended in this large-scale manipulation of the landscape, why do it? Our struggle to survive can be assuaged. When these ideas are built into material forms, their regulative power is reinforced. A concept is so much more real when we can touch it. Its certainty is guaranteed, and other options fade away. We stop thinking of the new environment as an extension of anything and simply take it as a prerequisite to our continued existence. With the bulb on a switch, we *assume* that we can have light and darkness whenever we choose to, independent of the rise and fall of the sun. Light and dark are found in a natural rhythm, but we wrench them into our own rhythm now that we have artificial light. We are able to forget the old cycle if it gets in the way.

We live for the most part these days within rectilinear structures that shelter us from the varying threats to our comfort: heat, cold, wind, rain, snow, and enemies who do not possess the keys to our doors. Do we build rectangular and cubical things because right angles have been around since the dawn of time? Husserl demonstrated above how Euclidean geometry is a consequence of the structures we have built for our extension into the lifeworld. (More on this in chapter 6.) Making use of the theory, we imagine that it describes a regularity inherent in the world itself. But what is explained is the universe of planned human objects. Euclid explains laws abstracted from a built place enforced out of right angles. We then have a theory that explains how we can build things, and we analogize, imagining that it applies to all of nature, or all of experience. But the extension of a desire to build need not pretend to explain that which we know not how to construct.

So what we build, in its apparent independence, returns to deeply affect the way we see the world. When we take what we have made for granted, the constructed environment stares right back, reminding us that we know only what we have put there and see what we want to find.

EXTENSIONS OF THOUGHT

Improving the Senses

Perception is an activity of the mind through the body, not tangibly changing the world, but assessing it through the input of the senses. When technology enhances these, the sense itself is transformed the moment its augmentation is possible. Perfect eyesight is no longer a rare trait once corrective lenses are available. The telescope and the microscope make our reach greater and more precise respectively.

Perception-extending technology does not alter the biologically given range of our senses. We still see only those light frequencies to which the eye happens to be sensitive and hear only those frequencies which the ear can pick up. What may be enhanced is the sensitivity itself; movements too distant or too tiny in the perceptual field may be intensified so that they seem closer to the human scale. It is remarkable how near planets may seem in the telescope, how many stars can be made out in the tiny circular field brought to us by the lens and the mirror. It is startling how much movement can be glimpsed in a tiny drop of water on the microscope's slide. Suddenly seeing these things, we know not what to do with them. Through the eyepiece of the two scopes, we have little power to alter the world we see. After turning away from these devices we retain some kind of direct belief in the reality of the worlds they present to us. We are more inclined to trust in the validity of micro- and macroscopic worlds previously known only through hearsay or authority. Once seen through our own eyes, such worlds appear accessible to humanity.

What has been seen passes over into what may be known. Whenever we see a drop of water, we remember how much may live inside us, at a different scale of life than our bodies can feel. When snowflakes catch on our coats as they fall from the sky towards the Earth, we spot the rudiments of their symmetry as they melt away, but we know the full extent of their crystalline beauty only after seeing them under a magnifying glass. If we have just seen photos of their myriad forms, the diversity is not quite as real, because it was not apprehended through a direct extension of the senses.

Visual acuity may be enhanced to bring in the infinite and the infinitesimal, within certain clear limits. After a point we must resort to more independent kinds of scale-changing instruments, such as radio telescopes and electron microscopes, which synthesize images out of the information they independently take in, without our direct apprehension. But we trust the mechanisms that guide them, accepting the results which they offer, further extending our range in the form of information provided by the machine, left for us to interpret.

In the aural realm, we may hear sounds far in the distance with the aid of amplifying microphones. Telephone and radio extend our sense of hearing. As they take in sounds from even greater distances, they may also be considered direct extensions of auditory perception. Our hearing is less important than sight in terms of assessing the world as a whole, but more useful than seeing in terms of interacting with this world spontaneously, as we speak rather than emit any form of light. Speech is received as a specific signal, while imagery is apprehended as context and whole. This may help to explain why there have been so many more technical enhancements of vision than of hearing. And our focus on the visual is as much a consequence of this technical direction as it is a cause.

Other senses? There has not been much need to enhance our sense of smell, except perhaps by compensating for its insensitivity by making the smells stronger, as in the use of perfume to intensify identification of people, places, or ineffable possibilities. The sense of touch may be one that is most enhanced by *lack* of intervening technology, which replaces direct bodily contact by cloth and growth of "personal space"—

walls we build between each other to hide the simple faces that connect us all. Yet the sense of touch is extended when we are able to feel close to other people and contexts that are quite far away. Some claim electronic communication will offer the closest connection between minds in the near future. But will it ever challenge the close touch of the hands alone?

If radio, television, wires, and satellites connect us to other people and processes in a visceral way, then we have extended the sense of touch to embrace other senses. The notion of *communication* thus begins to merge the individual autonomy of the senses. Linking ourselves to others upon the assimilation of *information* requires another movement away from distinction: perception must be merged with conception as we consider ways the processes of our thought may be changed through the building of structures that hold ideas together, giving them form.

The Tools of Abstraction

Thought is *abstract* when specifically separated from experience, independently represented and retained. An abstract thought is not always a sign of a concrete action, and cannot entirely be subsumed by an act of making. Retaining the abstract requires technology, albeit of a more ephemeral kind. The tools of thought, the techniques of formalized knowledge—these mold praxis into a form by which it may be assembled and passed on to others independent from what we may move with it or what we have seen.

Take the word *logos* to mean a systematic, regular order of any kind. I am saying that any constructive use of a *logos* is a kind of *techne* if something new is launched out of the process. This is a real stretch for the limits of the concept of technology, as it considers even what we say about the world to be a "making" of the world, not just a description. Through our techniques we come to know the world. Whether we are also able to transcend these techniques and learn *more* than what we are able to construct is the running question: can what begins with the human ever escape human limits?

Language is the primary technique of the abstract. Its rules and structure become detached from what it wants to represent as soon as they are set in place. Words are a step back from, but do not deny, the immediate. Call the technical part of language the aspect which is governed by rule. Then the most technological of languages are those which are completely regimented, wherein the information transmitted can hardly be misinterpreted. Computer languages offer instructions designed for machines. But real community-constituting languages are always also carriers of raw emotion, feeling, and wonder, so they are never only technological. They are so important and so successful because it is possible to say more than what is enumerated in words, drawing on a vast storehouse of allusions that enhance the range of what one person may evoke to another. Languages are the oldest of abstract technologies, those which have taken the longest time to develop, approaching nature's pace and style in evolution. The specific components of any given language seem to the neophyte to be a curious assemblage of rules, cases, and exceptions. Yet when we are accustomed to the language, all its curiosities seem to make sense in a *natural* way like the features of an animal: peculiar but essential in distinguishing the particular species from within the vast sphere of possibilities.

Languages are the least directed of technologies because they have taken so long to unfold themselves. The richness of actual meanings they reveal far extends their specific ability to transmit codified thought. A language is no simple logic imposed upon making. Language extends our experience by abstracting it, and then extends our formalized abstractions toward other listeners through the action of communication. More than information is sent out, and this is why ordinary language is never reducible to pure logic. But specialized languages which may be fully abstracted away from experience to work without any reference to their inspiration possess a different kind of strong unity. Symbolic logic and mathematics may be developed and communicated without claiming to represent experience, because one may speak and write them without any reference to perception or external action. In this sense they are the *purest* of techniques, because they make only ideas.

This is not to suggest that pure systems of thought have no influence outside themselves. Because they are technologies that extend our capacity to abstract from experience, they augment our immediate cognitive faculties. Technology always implies construction along with order. The invention of Arabic numerals, and in particular the placeholder of zero, made possible a great advance in the completion of complicated calculations, particularly multiplication as a process which need no longer be conceived as a long series of recurring additions. The failure of the Roman Empire to achieve a sustainable industrial culture may have had something to do with the clumsiness of its number system.[3]

But prowess in decimal crunching is not the same as mathematical innovation. The Arabs did not go after an industrial revolution either. Their conceptual genius needed first to be linked to human intention to build and to shape things into a hierarchy that could embrace all of society. The medieval church and state in Europe needed mathematics to build up the strength of its physical infrastructure. They saw clearly how the quantification imported from the Near East could change the form of their environment and fortify them against enemies. The practical application of pure abstraction has only increased a thousandfold since then.

Using pure mathematics to make better machines is not the same as building machines to extend our ability to think and work abstractly. A long litany of inventions has tried to improve our ability to perform regular calculations—from the abacus and the slide rule to the pocket calculator and digital computer. Such devices operate at a very general level upon the same principle: Beginning with a human cognitive process which can be codified, they present a mechanical analogy that garners the same results as the human mind, but by a method more appropriate for the setting of the machine. When we add and subtract in our heads, we do not have a series of beads rising and falling on rods, nor do we have two scales sliding against one another comparing and registering marks when we multiply. Do we instead have a whole series of switches turning on and off combining to find the value for complex mathematical formulae? Some researchers think we have neurons that work in this way, but we certainly do not *conceive* of mathematical problem solving like this. Unless, however, we have spent so much time learning how machines must translate problems that we forget that there is any other way to think of them.

Machines that extend cognitive dexterity independently of our guiding movements reinforce just those aspects of our thought which can be precisely codified into terms which a machine may understand. Digital computers begin with the simplest binary kind of mathematics, easiest to store and transmit, and may be used to model all kinds of complex processes through an intricate set of repetitive procedures. The more we extend ourselves through use of these devices, the more we tend to ignore qualities which cannot be represented within their constraints. This is not a necessary consequence of involvement with technologies of formalized abstraction, but at least a latent danger, always there when we forget that we can sense more than we are able to make and describe. Chapter 4 will investigate the powerful effect of machines that resemble the mind on the way we speak and imagine the world. But a simpler side of technology promises a more immediate power: extending our capacity to retain what we have experienced, without abstracting it through a formalized language of any kind. Is this possible?

Material Memory

How much experience can a person retain? How many concurrent streams of activity can anyone keep track of? It is commonplace to suggest that each of us utilizes only a tiny percentage of our brainpower. It is also widely believed that those not exposed to the technology of the written word are able to remember long oral accounts without too much difficulty. Those of us in a visually literate and document-oriented culture place far more stock on the accumulation of information in storage than on any individual's unaided ability to retain it at the tip of their tongue. When we need to know a fact, we look it up. The collective memory of our society has taken material form. Volumes and volumes of structured information await our perusal. We may read them, listen to them, watch them, or plug into them from any desktop terminal.

The world of information requires language, so it already makes use of a certain amount of abstraction. However, there are certain technologies that claim to extend experience simply by capturing it, freezing it in a form we can recall whenever we choose. The Greeks thought that *techne*, as art, was largely mimetic, imitating nature and experience in a manner which was, at least to Plato, inferior to the real thing. Subsequent concentration on art as creation rather than reproduction led commentators to identify the mimetic side as an imitation of the operation and process of nature, not its appearance to us through the senses. This seemed adequate until the development of techniques that were mimetic to a far greater degree than previously thought possible: photography and sound recording.

By calling these and related tools material extensions of memory, I am suggesting that they play directly on the processes by which our naked minds hold onto information. They deserve to be separated from those devices which function by specifically formalizing abstraction. But I would not go so far as Levinson and suggest that these are means to transmit "representations of external reality with little or no abstraction."[4] It is only that the method of their abstraction is more subtle, appearing to embody pure experience.

It is a cliché that the photograph can never lie. Even a casual snapshot is the result of an intended perception. The pointer of the camera has decided what to attend to. The time of exposure is, with present equipment, far shorter than the duration necessary for a human perceiver to sense or remember. The image is fixed and static, and as a photograph it seems to breathe an authenticity which no painting can possess. However faded the crispness of the emulsion, we still remember: a machine has captured this. That pattern of light rays was actually there, frozen in an interaction between the human act of inciting the machine and the machine's predictable response. The camera does not only augment our records of the external, luminescent world—it extends our awareness of the intent behind perception. Are all green peppers more sensuous once we have seen Edward Weston's seductive photograph of one? Yes, as we realize how the vegetable is changed by the way we choose to see it:

> This was not because the camera lied, or even because it told the truth, but merely because it saw what it could, not what it should. The camera could be as immune to beauty as it was to banality. Subject matter wasn't being duplicated or interpreted here, *it was being transformed.*[5]

Photography as technique, then, transforms visual impression much like any other method of recording the senses: it has its own rules and principles into which it must squeeze experience. It abstracts from sensation in many ways—stopping time, aligning the image into a frame straight from the rectangular world, flattening the always-felt realism of depth in constant movement. Yet, in contrast to communication in the throes of symbolic discourse, we do not need to be aware of all this to *use* photography. So it quietly changes the way we remember events, people, and places.

Returning from a journey, our friends ask to see photos, slides, pictures of us in the thick of the strange, distant place. This is somehow proof that we have been there, windows of real experience which they can connect to. But they are small, bounded, and narrow windows that already abstract from any living happening by halting its motion. Even video or film reports of another person's experience are still the account of a gaze, the flat detached story of a moving series of observing intentions. Every attempt to reproduce an experience is consciously severed from the original location of the experience, so it will never be a totally concrete version of it. Experience cannot repeat itself. Any recall will be somewhat faded.

Photography claims more realism because we believe the behavior of chemicals on an emulsion plate to be free from bias. As a tool to enhance our memory of the parts of our own, personal worlds, it enjoyed its first surge of popularity through the rendering of portraits of family members near and dear. With photographs, we can carry reminders of the people intimately connected to our lives, even though we may be physically detached from them. But not all photographic reminders are revered as accurate. When a person is dissatisfied with the quality exuded by a photo of themselves or another, does it mean their appreciation of the subject disagrees from the truth revealed in pure representation by an unbiased machine?

Here is a glaring example of how a technology transforms not only the objects it creates, but also our notions of correct depiction and memory. We have all seen photographs that seem to misrepresent our own human retentions of their frozen subjects. This use of photography demands the application of its own criteria of detached, unequivocal observation to our own minds and inner retained histories. We applaud the rare human capacity for "photographic memory": recalling the past as a pure and detailed image, with no fudging of specifics through personal bias and choice regarding which features are important enough to be retained over time. Doesn't this notion of a better kind of memory lose something of the rich and diverse range of possible human glimpses of an incident which can make it a wholly different moment for each person involved? (Who wants a past that is only a path of unforgettable glances?)

So separate it from memory—consider the photograph as art. It capitalizes on the unabashed record of strange, beautiful ways of looking, pointing, and shooting. Call the result art and you praise the intentions more than the process. Still, the photograph is a technical result, a thing made more directly out of light and shadow than any painting. Someone did see the frame in a viewfinder and eventually shaped it into the finished glossy print; the root of a photograph remains a tiny flash of a human visual grasp.

And what of photography as a substitute for our memories? I think back on the places I have been, and on the persons I have been. I remember only scattered experiences of my early childhood, and these seem distinct enough from the collection of material, yellowing images of myself from the same period, taken by others, those who were watching my progress and f-stopping it through the lens. There are recollections of mountains climbed and faces kissed. There exist photographs of the same. Which past will refuse to age? It is most disturbing when the uncaptured memories seem to take the form of photographic images, static and frozen, even though it is not through the aid of photographs that I have been able to remember them. My memory, as I delve farther and deeper into it, seems less like photography in mimetic accuracy but more like it in the limitation of acceptable image: singular, sudden, and trapped.

And now here is a reverie on the subject documented in words! What will I remember of it twenty years from now—the thought process that led to it, or the photolike picture of myself sitting at the computer, staring either at the screen or off into space? The point is that language and photography prove to be distinct ways to abstract experience, each retaining some of the dynamism of the happening, but exchanging another part for abstract reproducibility in order to remake the lived event into something recountable for others. A story might retain the temporal rhythms, an image visual rhythms, but when each is redone, the original source intentions and perceptions are remade anew.

Machines of memory are designed to record only certain possibilities. Photography takes as its possible data only the infinite series of possible rectangular images which it is able to immortalize on film, guided by our glance through the lens. Sound recording devices substitute microphones for our ears. Although we may guide their accuracy while wearing headphones, the way we use our hearing leads us to listen for different criteria. We cannot use hearing to construct a disembodied vision of the arrangement of objects in space; only sight can be abstracted to imagine this. We hear the array of sounds in relation to our position; each noise reinforces our location and context. Because we are not so easily tempted to abstract from what we hear, the recorded voice is often closer to memory than a recorded image, probably more effective without a visual element to "enhance" it. (So films and television leave less to the imagination than radio.) If we only hear the voice of someone who has died or is far away, the remaining part of our memories compensates for the lack of information. The limited extent of the mimesis in sound shocks us into recognition. We expect to wheel around and find the absent person sitting right there, speaking and reacting to us as we are best able to remember them.

Because our aural sense is less sensitive to imitations, and perhaps less dependent on reinforcement from other senses, we are able to enrich memory through the tools of recording. It is not a tangible substitute for the experience which it represents, but more clearly a repetition of the aurally sensible aspects. One may learn to read music from the printed page, and then experience difficulty learning to memorize after too much facility in the re-creation of sound from the written notes. Learning to imitate music by listening to recordings requires more immediate attention to the temporal pattern of the original, as the sound has not been re-presented in the form of a systematic symbolic code, but only as changing sound. So learning *by ear* as opposed to *by sight* often leads more directly to an internal command of the stream of sounds and a knowledge of the music equivalent with the ability to make it.

Of course, using the reproduction of sound as an artistic medium is different than using its directness to spur the internalization of experience. In its more sophisticated forms, sound recording onto cylinders, disks, tapes, and digital media is capable of as much deception as photography. A human voice can be equalized, chorused, dropped down a few octaves, or sped up so that it becomes a far cry from an extension and instigator of the memory of its personal source. The spatial characteristics of a recording can be completely renovated through the illusion-producing technology of synthesized reverberation. But these uses leave the enhancement of memory far behind. Starting out as a tool to record heard experience, the tape recorder becomes an instrument in the creation of works of art. As one of very few nearly mimetic technologies, it grasps some part, never the whole, of the sensual inflection of the world that surrounds us. Just the ability to accomplish this at last reinforces the certain stance of this world toward us. There is always something to be heard if we choose to listen.

The retention of memory soon blurs with the abstraction away from experience necessary to efficiently record and then process information. Computers also retain huge quantities of information for us, but only if the information has been successively transmuted into the digital realm. What is stored there has already been broken down. Boundaries between these last two categories soon dissolve. This is where technology is going—combining its features to elude demarcation. Environments will think, we will work together not occupying any common physical space, and our meaning will inhabit an ineffable sphere called virtual reality, only real in the link between hand and screen, touch and computation. This will be the final challenge for chapter 4.

THE TEST

Here are six general ways that technology enhances experience, divided into two basic groups of three: the extensions of action, and the extensions of thought. Such an eminently dissolvable dichotomy might appear contrary to the dynamic, searching spirit of this work. But I feel compelled to divide technology up, to prove it can all be seen as an extension of the human drive. This scheme of categories needs to be tested, as a series of *tendencies* which might help clarify the ways techniques differ from one another and the ways in which technology as a whole differs from and relates to humanity.

First Dichotomy: Action or Thought. The former category refers to all those technologies which work practically to alter something about our material world by changing the ways we physically relate to this world. It is not as if action is wholly other than thought, but only that these tools transform our environment. Those technologies that operate primarily on thought do not have these immediate material effects. They change instead the way we take in perceptions of the world, the way we conceive and manipulate these perceptions, and the way we are able to retain and recall experience. Yet they are by no means wholly conceptual; when they are put into practice to build the external world in a new way, they inspire and combine with techniques of the first kind.

Second Dichotomy: Means or Construction. The most direct kind of technology is an immediate means for realizing physical or mental intention by extending the forces of the body and mind. The shovel transfers our moving weight down into the earth; the magnifying glass enlarges the moth's antennae without significantly challenging the notion of what it means to see. These techniques reconsider the range of these basic gestures. Constructed technologies, in contrast, are more sophisticated as self-contained systems, and their use requires an alteration of basic intentions. An abacus or calculator requires new ways of thinking about number to input to the machine. The way we conceive of number is immediately changed. Driving a car, we do not need to be completely cognizant of the way the machine works, but only how it enables us to get from point A to point B, regularly, consistently. Our interaction with the landscape and with distance is altered at once.

Third Dichotomy: Driven or Embodied. The more immediate kinds of technology require a human guide in order to function. The boomerang does not fly unless someone throws it, and the path of its flight is a direct result of humanly applied force interacting with the design. It is the same with a complicated construction like the jet airplane, which will never fly without human pilots guiding it, although the energy needed to get it off the ground comes not from our own body but from constructed machinery. Both of these technologies are driven directly by human intent. It is a leap from here to the alternative class, which, although instigated by humans, functions independently of the initial actions that created them.

I am speaking here about technologies that embody their purpose, extending human intention by rendering it tangible and self-determining, placing it into our environs, so that when we assess the world around us, *we perceive a world already transformed by ourselves*, so that any separate "nature" must be abstracted from the technically immediate world. Roads, houses, telephone lines, and irrigation ditches are all part of this type of technology which embodies our need to communicate with one another and shelter ourselves from adverse natural conditions. We look at a street of detached dwellings, we see territorial areas. With the wires connecting them, we see the potential for interaction. Sure, these technologies may require continual maintenance, but they function without requiring continuous control by an intending person.

Technologies that constitute our environment, forming our material infrastructure—these are the clearest examples of embodied techniques in the realm of human action. But what of thought? Can there be an embodied technique there, independent of our guiding intent? Certainly this would be impossible without blurring somewhat the distinction between thought and action, because any embodiment must to some extent be distinct from the human mind. It must have a material component, and not be a purely mental technique like mathematics or Plato's Ideas. It need not be independent of human intelligence, just able to enact thought processes without constant prodding.

The material extensions of memory are the simplest examples, but we have shown above that their material presence is only part of the story. Their use may require as much abstraction and systematic apprehending of symbols as any other transformation of mental processes. There is the world of organization abstracted by bureaucracy—this is the ideological infrastructure where form follows only function. Or consider the computer, with its flexible "user environments." It is not appropriate here to ask if it can think *for* us. The question is whether any conceivable structure can embody the way we think, making the operation of our mind somehow tangible in our conceptual surroundings, independent of any particular intended thought process. In chapter 4, computers will be shown to claim to offer a world and a context. But it is never the whole process of the mind which is mirrored, but again only those selective parts which may be systematized, then analogized, in mechanical form. It is when we mistake the part for the whole that our intentions are unsuspectedly altered.

With these three dichotomies imposed upon our original six exploratory categories, here is a diagram that tries to represent them all at once:

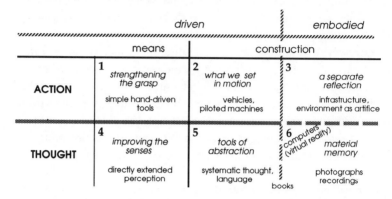

Figure 2. Tendencies of technology

Any graphic representation like this makes such analytic categories appear far more lucid and discrete than they could be in practice. The goal here is to recognize certain *patterns* of extension which course through different instances of real technology with varying amounts of blending and distinction. The dotted lines suggest an area where two tendencies combine to articulate the complicated, many-layered technical extensions of the ambiguous human edge. New developments threaten to blow these categories apart, though still letting earlier kinds of tools continue to exist. And the future abstract machines will doubtless contain clues of their roots, conventions left over from the old human constraints.

Just as significant is the identification of those human attributes which are *not* extended by technology. Some of the most revealing qualities in the pinpointing of human essence have not been improved by artificial advance. This may explain why it is possible for technology to extend action, need, perception, abstraction, and memory at all.

THE HUMANITY WHICH REMAINS

Technology makes us physically stronger as we build with it, perceptually sharper as we see through it, more facile with concepts as we think with it. Does it help us make better moral judgments, further our sense of place in the universe, or enable us to love one another and the Earth more completely? Many would be quick to answer "no!" Yet we want to believe that technology somehow improves human life, else why would there be so much drive to change our lives with it? Techniques realize our intentions, which should be part of a deeper sense of purpose if they are to genuinely improve humanity.

Consider those roots of intent which seem *not* to be enhanced by technology. It is no accident that the faculties of judgment, moral reasoning, and spirituality are the very attributes most often used by philosophers to separate humans from other living species. While the archaeologist might emphasize tool use or food sharing as those qualities which set humanity apart, philosophers are more apt to point to those grayest areas where they claim the greatest expertise. These clearly important components of human life are also those most difficult to describe in systematic terms. And they cannot be enhanced in a directly technical manner as perception or physical strength might be. But as they might profit from the handling of more information, they are involved with the kinds of technology that make more data available. They may be challenged by technique whenever the machine is put forth as an arbiter on the side of unequivocal truth. The more we value exactness, which is primarily a consequence of technology, the less we are able to respect the elusive and exclusive parts of human decision and propriety.

Observe how these faculties get involved with technology: Judgment comes into play whenever we decide between two or more alternatives, picking the best or most correct one according to a mix of specified and unspecified criteria. If our choice may be made clearer by the possession of unbiased information independent of human-intended observation, then judgment may be enhanced with technology. An immediate example is picking the winner of a close race. A whole box full of judges can argue over who they saw come in first, but if we have a slow-motion video recording of the finish, we can study it frame by frame to discover who crossed the line before all others, with far finer resolution than that of any human observer. If there is *still* an argument, we may only resort to our own abilities of persuasion, but the technological record is usually regarded as the neutral observation. The camera cannot tamper with its own memories of the event, most of us naively believe. As a witness, it is not supposed to lie, so we unsteady humans defer to it in cases where it is conceivable that the truth of the matter is something which might be observed, but not by an unaided human eye.

In legal judgment, notice the similar powers of material extensions of memory. Photographs and tape recordings made through "bugs" are often presented to the jury as evidence far more objective than anything a human witness might tell of an incident. Their strength and perceived honesty is so great that the opposition has little chance to rebuff them, unless they were obtained through illegal means. We have such a law to protect the privacy of our actions and words, not the least because we know that manipulative objectification of these acts may certainly misconstrue them—promising truth within snapshots of reality through devices that are unable to assess the situations they appear to represent. Why then do we trust them so?

Out of fear of an irrationality latent in human decisions. There seems to be more likelihood of making a mistake if we are unable to precisely quantify the reasons for any particular choice. How can we answer our critics if we are shown later to have made an incorrect decision with the discovery of more information? Will they accept the answer that we simply followed our instinct, intuition, or intention? In the end, technology never decides anything for itself; it is we who make all the decisions, and in this sense our judgment is always independent from technology, not enhanced by it, but only provided with more data which may shake the faith in our own sensibility. Technological sophistication does not always provide an unbiased outlook on a puzzling situation. It is also quick to complicate things. New medical machines may extend the lives of a few lucky patients, but their scarcity makes them unavailable to most. How are we to decide who gets a chance at them? The technology is absolutely no help here. We are forced either to choose randomly among the qualified patients, or make some kind of cautious assessment as to whose life is most worth extending. And this requires developing a notion of comparative value of human life, a thankless, troubling task. The machine cannot help us place a value on ourselves.

So technology may help us judge by claiming to provide a dispassionate account of a disputed event, and it may create new unsolved dilemmas by changing our notions of what may be done to solve a problem. But there remain important areas where technology is *indifferent*. We have never been able to extend acuity in moral reasoning through new techniques. We believe it is wrong to murder whether the weapon is a knife or a pistol. To the extent that these are moral statements, they are unchanged as they involve different methods and different objects. Yet the moment they are construed as actions, they once again are transformed by technology. Knifing someone is a far more difficult and physical act than shooting the same person from a safe distance. Pressing a button to start a war can be done with immense personal detachment. Taking money may seem abstract and symbolic enough not to worry many would-be criminals, while stealing pets is not so common—the thief often leaves the personal behind. With its distancing and abstraction, technology may not aid morality by helping us see choices more clearly, but it makes action more visible than responsibility.

Does this make morality a human attribute which needs to be protected against the dulling, equalizing tendency of technology? One response is to affirm its independence from action and announce a moral imperative to use techniques in a way which is consistent with independent, nontechnical notions of the correct and right way to act. But this too is naive, as we have demonstrated that all action is inextricably involved with tools and instruments as long as it mediates between a human desire and an effect on the world. It is not enough to protect morality from technical indifference, but we must carefully consider the ways technology challenges morality as the circle of technical influence comes back on the instigating wish. Although moral thinking may not be enhanced by any existent kind of tool, it is certainly affected by the steady change of choices made among the field of possible actions.

And how do we recognize our place in the throes of the universe? Religion has traditionally provided this, as we cast our fate in with a faith, adopting a system that explains precisely why we are here now and what will happen to us after we die. Science and technology may certainly carry on within religious belief, but as they provide new information, they encourage the questioning of the dogmas that define the whole. The system can take only so much of this before it is upset, as Copernicus and Galileo learnt well. As we concurrently learn to do more to the world, our own hubris grows with what we accomplish. We *see* the sun as the center of the solar system, but *act* as if we are the center of the usable universe. We still look at the world in wonder, but live in the world we make. As nature comes to be conceived in praxis as a means to ensure our place, the stature of this place rises. Wonder at what we can do with the resources we discover begins to outshadow any humility in the face of immensities off the track of our purposes.

Still, the enveloping reach of use does not provide the reason for use. Human purpose may only be intensified through technical action, never replaced or fully represented. We do not succeed in dissolving Pascal's two infinities simply by expanding our sensitivities toward the macrocosm and the microcosm. The unknown remains at either end, even as we guess tentatively at the limits of our ability to conceive of scale.

Again, technology heightens wonder whenever our awe of the universe may be encouraged by information. When we watch films of the lives of strange creatures in the deserts and jungles which we could never witness in person (because they are both far away from us and perhaps active only at night, when only the camera may see them), reverence for the intricacies of life reaches beyond immediate experience. This affirms our sense of location in the environment only if this sense expresses an interest in other forms of life, leading to empathy with non-human ways of being. This is not a sentimental move, but a step toward spiritual recognition of humanity contained within an encompassing universe, connected to all the things we may learn about. This would be a religion which could fortify itself with information, and would not let itself be destroyed by the arrival of new facts with time.

Here, religion itself can be conceived as a technology to the extent that it puts abstract and objective information into some *pattern* that engages us. This may be dangerously premature, as knowledge of one's proper niche in the cosmos is nothing which can be rigorously defined or organized. It should never be abstract. This feeling is inextricable from experience. We attach it to "spirit" because it is not encompassed by our notion of reason. At the moment it touches us, it is not a technology—*here* a line may be drawn.

Yet if we are to realize our spiritual flash of recognition in the material world, we do connect belief to technology. The criterion of *self-realization* is often invoked to make the feeling of rightness in the wake of the world traditionally provided by a religious belonging into an active goal which individuals may choose to seek. It is thus a distant descendant of Aristotelian notions of humans being on Earth to fulfill our potentials, to be all that we can be. This involves technology at the point where an attitude is defined toward the surrounding world. A key component of psychologist Abraham Maslow's list of criteria for recognizing self-realization is "independence from the environment."[6] This suggests a goal of being where one is not bothered by the external whims of the world, where we are our own masters and masters of our context as well. It is a viewpoint centered around an individual's psychological self-realization, nothing larger. Maslow relegates the environment to the role of insurgent distraction, something which worries us less and less the more we are in control of our potential. As this kind of self-realization progresses, we gain personal confidence and come to let nothing outside us faze us.

This idea of self-realization is apt to value only those tools which free us *from* the Earth. It is built upon a very narrow notion of the self in opposition to the environment. In contrast to Maslow, Arne Naess presents a vision of Self-realization with a capital *S* to signify an embracing of both the individual self *and* an all-encompassing Self of the universe, akin to the *atman* of Hindu philosophy, yet without denigrating the value of the individual self in the pursuit of the whole.[7] This metaphor is introduced as a tool to encourage identification between our own realization and the self-fulfillment of all of nature. The more we understand ourselves, the more we understand the world. As the approach is developed, we retain a certain autonomy, but only as much as we *comprehend* our place as part of nature, not in distinction from it.

This image of the universe as being enough like us to share the name of "Self" appears whenever the knowledge of how to make something is reconfigured as knowledge of how the universe works. It is present in the bow and lyre of Heraclitus, in the potter's wheel of Plato, and in the mathematical "book of nature" read by Galileo. When we wish to identify with the world, we envisage it as somehow like us, as chapter 4 will show. The processes revealed and controlled by technology are a powerful part of this likeness. But is it more than likeness? Technology certainly helps us to reach those aspects of the realization which may be quantified and materially constructed. This in turn helps us see natural processes as things that function according to patterns similar to those which we build ourselves. Can we say that technology reveals possibilities for the Self-realization of nature? This is a great leap from faith in human understanding and human progress to a claim to know what's best for the world at large. Here is trust in the power of unbiased truth far beyond the disinterested flash of an arbitrating camera at the scene of the crime.

Few philosophers will still claim today that the purposes of humanity are equivalent to the goals of the universe. And yet so many of us want to believe that we remain the center, striving to fulfill a divine mandate to reveal the cosmos in its fullest meaning. This holds philosophy and/or religion to be far more than techniques whose function is to establish a sense of place in the horizon of experience. No—if we are so sure of ourselves as to be sure of our images of the universe, then we put ourselves forth as the embodiment of truth.

The conclusion in chapter 6 will be that it is wrong to push technology and its metaphors for nature this far. In all the examples suggested in this section, *technology never replaces the faculties which it challenges in its drive to extend.* Judgment, morality, and the sense of place must be part of the remaining humanity which is invoked whenever a technique is intended in one direction or another. At the other end is the region towards which technology is extended, in which our intentions are contrasted, as idea and then as embodied realization. These are the *limits of nature* whose various conceptions will be discussed in the next chapter. And we cannot claim to know the full content of these limits, as they would no longer function as borders if dissolved through total comprehension.

It is also wrong, however, to imagine that technology pushes us *away* from the most essential parts of our humanity. Through embodiment, practical realization, and the returning arc of influence, human dynamism is expressed through technology. Every tool contains some of the spirit of the artistic conduit, like the pencil or the clarinet. A device that can turn the rough motions of the hands into precise and variable gradations of line and shading opens up desires which we never had before. Similarly, a musical instrument that offers a key or a tone hole for each finger of our hands allows us to choose notes faster than we can think of what should be played. Whole musical phrases well out of us into the material realm of sound before we can analyze what we are doing. While making art, we are anything but independent of our environment, expressing something of ourselves through material form, engaging the world as we are startled by what comes from inside us.

When we use the word "art" to describe an act of human creation, we usually mean there is a part of the object made which is independent of its utility, its function, or its realization of a clear human intention. Something else is expressed, a quality is revealed, an emotion is conveyed, beyond an ability to explain or categorize. Art is part of the original meaning of *techne*. A component of artistry should still be retained, both in the beguiling way *more* than the precise intention comes to fruition through the machine, and also in the way we apprehend the change which technology wrings upon the precise intentions that put it into use.

How do the different kinds of technology affect the boundaries of our vision of what we are? The beginnings of an answer may be shaped around the categories introduced in our table of types of technology as extension. Each category presents an affront to what we first think are clear human limits. Although the techniques initially push back these limits, each poses questions of essence as much as it opens up new possibilities for action. How does the transcendence of our limit of physical strength affect our sense of invincibility? When our environment is made more amenable, do we feel external nature is more like us or less?

And in the possibilities for extended thought, how do we assess the idea of human "scale" when so much more of the universe becomes perceivable? How far do we move away from pure survival when communication becomes possible? And, finally, when memory and thought processes are mirrored in material, what happens to our bodily limits in place and time; might this be the only tangible way we could seem to be eternal? Here are some of the toughest challenges to our sense of self:

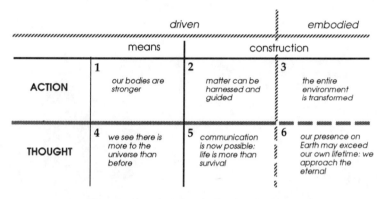

Figure 3. What each kind of technology extends

Where there is extension, there is the danger of loss. The next task will be to examine the use of the concept "nature" as a possible check on the transformative power of these challenges, and then to clarify the distinction between what is intended and what is expressed by unfolding a vision of technology which begins and ends in nature, changing its definition in the thick of a process that is both thought and action. *The firm grasp of the hand soon lets go in favor of a lighter touch: a button is pressed, the machine is turned on, and we can only hope it makes sense to the landscape.*

Here is a gently harmonious machine, with the blocks and pistons appearing to flow out of the swamp and into the river. The wheel steps many times in the water, turning the crank to dredge water out of the soggy earth. It is a machine propelled by natural forces, set into the landscape. Look at the symmetry with which the picture is drawn: the line of the garden, the nine planted trees. A gentle plan in the quadrangle eased out of nature. Although we might gasp at the draining of a precious wetland today, this was one of the prime processes once used to prepare land on the water for transformation into the great seats of civilization. The human and solid place is refined from the impenetrable swamp. Pushed by the hand of flowing water, the dry land we desire is soon squeezed into being.

This machine looks like it fits into nature. In this chapter I wish to argue that theories of technique throughout history have presented this fit as a common goal. Any definition of technology implies a definition of nature. Technology takes a stand on nature if it is thought to transform any existing order, or to reveal it through analogy. Manipulation of the environment is *the* deciding issue in the controversy of how far humanity is included within nature, and how far we are a contrast to it. The difficulty in establishing such a relationship seems inherent in the multiple meanings the word *nature* has tried to assume.[1] If nature is a synonym for everything contained in the universe, then it carries little normative weight. If it is intended to refer to all things and processes *outside* of human alteration, then we have little hope of finding any place within it. If technology contradicts it, we can do nothing to improve our "naturally" given place in the world. If technology represents it, then the world is never more than what we can know of it.

For nature to be a guide, it must be the carrot at the end of the stick—always tasty, always out of reach.

We come from nature, then we decide what the word means. Which is it to be? To escape the quandary of the either-or, the following idea may be helpful: nature is both where we have come from and where we are going, while we are lost in proportion to our distance from it. By moving rapidly through the history of some of this speculation, the influence of technology on the articulation of this vision will be revealed. Nature begins as the grounding idea of a primarily fixed world, but as humanity defines itself more dynamically through historical awareness and change, nature will need to be more than the set and everpresent context for our developing techniques. We may only retain it as a goal if it is able to guide technology as well as encompass it. Nearly all proponents of technology want to approach or complete, if not to master nature. Seldom does the machine turn away from its setting. Yet the nature that we grope for moves as soon as we touch it.

ARISTOTLE SWIMS IN NATURE'S RIVER

At its root, the word *nature* never meant to refer to a static totality or a fixed order outside humanity. The Latin *natura* is derived from *nasci,* "to be born," originally used by cattle-breeders to describe the birth of calves.[2] As such it was a likely candidate to translate the Greek term *physis,* a more abstract term, but still primarily a mobile process, neither static event nor imprinted plan. *Physis* comes from *phyo,* "to put forth," originally describing how plants put forth their leaves, branches and flowers, and animals put forth fur, horns, and feet. *Physis* identifies the moving drive within things, and *natura* was applied by the Romans to refer to the same. So the "nature" which we now wish to apprehend as a totality and guarantee of order in our world may be traced back to a concept of movement, a growth that always comes out of something else, a transformation always tending toward a new form or goal. Can the *logos* of a moving physic be compatible with an enveloping nature that depicts the world and its limits?

This is why nature as fundamental movement inspires so much rich controversy. Once we analogize from the observations of birth and change in the biological world to the theory of a guiding whole, we get as many new questions as answers. Aristotle realized that by admitting the existence of *a* nature, one imagines some universal pattern holding the presence of life in distinction from chaos. When we say animals, plants, or waterfalls exist "by nature," it means they all contain within themselves the propensity to move and change, as well as the unifying fact that identifies them as objects. No thing which is conceived by nature is all growth and change; there must be something about it which ensures its singular identity.

Only natural objects, though, contain the ability to move within themselves, according to Aristotle. This distinguishes them from the mere products of *techne,* which are the result of human harnessing of natural process, in themselves static: no longer living material. Artificial products "do not contain the source of their own production."[3] On the other hand, *physis* remains more than a definition of life, for inorganic processes also demonstrate change in and among themselves. Human processes and directions were never to be excluded from it, only the inert aspect of the things we make. For Aristotle, the most salient meaning for nature is the *ability to come into being* found in all things deserving of the name. He also admits an understanding of nature as material, that out of which things are made, and, more pointedly, the notion of nature progressing toward a purposeful end. This also happens at the level of the individual natural entity, which completes its own nature if it reaches its final *entelechy,* actualizing its potential and fully realizing its own character and place in the overall *telos.* All parts of nature have their place and potential for fulfillment, from the uplifting mountain range to the human life striving to imbue itself with happiness.

The overlaps and conflicts enjoyed by *physis* in Aristotle remain with us whenever we try to extend the meanings of the word *nature* today. We like to see it as a primary material, out of which things may be made, and we also feel comfortable considering natural objects to be those which act upon their own volition. But why does all this action happen? Is it going anywhere in the long run? Several fundamental ideas have helped to clarify Aristotle's conception of a goal-directed *physis* since his time. The most support and challenge for his idea has come in the guise of the theory of evolution, which in its most general form posits that life forms have evolved from one into another, in ways which can be observed and hypothesized from past forms. Developments of the theory explain how the change has occurred, in steady movement or abrupt jumps. On the other hand, the theory of natural selection suggests that the tremendous diversity of life forms on Earth may have occurred *without* a guiding plan, simply appearing through improvement of adaptation and adjustment of living conditions after random mutations. Does this shoot down teleology altogether?

Not as much as the earlier revolution within philosophy, which succeeded in separating conception of mind from conception of material. The same Renaissance that made possible new scientific discovery thrust humanity outside the realm of the sensed and the accountable. We became first observers and inquisitors of the world, and then later users and transformers, as the meaning of material was recast as the service which it could render to us. No longer striving to fit into the world through our understanding, we instead use knowledge to remake nature in submission to *our* teleology, making it true only when it helps realize *our* goals.

And how clear are these goals as our transformation of the world accelerates? It might seem that the mechanism of change suggested by mutation and selection may well be enough to explain the vast technical alterations in the human condition which have occurred since the year 1400. Certainly no universally accepted purpose can be cited to justify the movement we call progress. Like Aristotle, we want to believe all motion starts somewhere and ends a definite somewhere else, part of a larger benign design. Others have called it the realization of human freedom, the course from nature to spirit through history, or the extension of nature through humanity towards a higher form. But the moment we try to explain the change wrought by humans on Earth by *distancing* ourselves from nature, we include the likelihood that such change will render us less sensitive to any independent purpose the world may be following on its own.

I hope that the growth in human self-consciousness of purpose does not require that we break away from the encompassing kind of nature within which Aristotle tried so carefully to elucidate our place. There should be a way we can accept the transformations and benefits brought by technology to aid in reaching a nature that includes us the more we learn about it.

Aristotle's *telos* in nature is not an idea that dies easily. It is with us whenever we speak of a correctness "by nature," whenever we suggest there is something unequivocally right in the natural world, worth preserving, respecting, or understanding in itself. We need to examine how technology changes our grasp of the particulars of such a nature, while changing our entire conception of it as our ability to act is strengthened and guided.

Nature needs then to be made into something which can provide a sense of correctness, as well as referring to the context out of which humanity struggles to define its place. Baruch Spinoza is important here because he brings the notion of God down to Earth, admitting that we come out of nature only so that we may learn how to know it intuitively, progressing toward interconnection with all individual things and the unity they share.

SPINOZA SEES END IN BEGINNING

Spinoza shows how we may conceive of human life as a clarion call toward the attainment of nature. How will technology play a part? For Spinoza tries to find a way to uphold the idea that mind and body are distinct alternatives, while not naming them separate substances. They are conceived as different attributes of the one substance, alternatively named God or Nature.[4] Humanity is consequently not set against nature, but comes out of nature only to strive to return to it through a uniquely human approach to perfection: directly apprehending the fact that all things hang together. He describes this direct but achieved intuition as a grasp of the truth, yet in practice, I believe he is suggesting a particular way of *living* towards nature, a guide for how to have the most human of lives. But these specifics are implicit in Spinoza, and we will have to provide them ourselves. I hope to show that this conception of nature at our beginning and at our end will be able to offer some help in the choice of those technologies which will be most appropriate for the search.

Spinoza asks the overall question "How should humanity fit into the world?" This is *the* ethical question, as it asks for an answer that would tell us how to live, implying that we have some choice in the matter, as we are not equivalent with the world. But as soon as the question seems to place a separation between us and the world, we learn that there is but one substance. Yet the one substance cannot just have one name: it is *Deus* sive *Natura*. Does this *or* admit an uncertainty? Not nearly as much as it implies a God that *is* the world, rather than a God that stands apart from the world.

Notice how awkward the presence of the *or* seems. Spinoza is drawing this uneasiness out of us, in an effort to conflate these two concepts into the single perfect idea that should reflect reality's order. We may have always known these words were related, but there is a set tradition delineating the distinct use of each term. If I look at each place in the text I have used the word *nature,* it would seem an awkward leap of faith to insert the term *God.* And to replace *God* with *Nature* would in most cases make our reverence too immediate, unabashedly *here.*

This is exactly the discomfort Spinoza wants us to wrestle with. And so he was named a heretic for speaking of a God who is singular only in the fact of being everywhere present. This becomes a God not to be worshiped but to be *understood,* where understanding comes to mean a knowledge whose certainty and tangibility transcend reason to be revealed in an absolute and unequivocally present intuition. To seek this kind of knowledge is to learn how to live. *Episteme* is now tied to *praxis.*

But even God in its unity is from the first divided into two aspects or manifestations. These are the medieval concepts of *natura naturans* and *natura naturata.* As two species of the genus Nature, Spinoza identifies the first as "that which is in itself and is conceived through itself," while the second refers to "everything which follows from the necessity of the nature of God."[5]

The words themselves suggest a division into processes (Nature naturing) and objects (Nature natured). In a dynamic world, the only aspects which are conceived within themselves are those which flow through the continuity of things, making things possible by their linking. The instances and particularities follow from the necessities of movement; the things we perceive as natural result from the unseen but everpresent forces behind them. What exists by virtue of its own nature alone is to be called *free,* while what is compelled by another in a fixed and prescribed manner is called *necessary,*[6] according to Spinoza's use of these terms.

So the *naturans* is akin to the instigating and the moving while the necessary *naturata* is determined by the free play of the first. If freedom is then the creative side of Nature, could there be any other kind of outcome than Nature-as-manifest? This depends on whether human freedom is entirely contained within natural freedom, or whether we might choose otherwise. Human intellect itself is then surmised to be part of the active Nature, but what it *refers* to is the passive Nature.[7] Because active Nature is the creative energy surging through us, we cannot escape it. Our freedom enters when we decide which parts of Nature to focus on. Spinoza's answer hinges on the fact that his notion of freedom does not reflect any lack of constraint, but a kind of *peace with oneself,* a state of sheer self-determination. The difficulty of achieving such a place will prove the toughest test of his principles; this double *Natura* seems to be one that both draws us towards itself and thrusts us away.

Humanity is distinguished as a part of the world because we think, and in the course of our thinking we form ideas. With this notion

Spinoza directly follows Descartes, while aiming to avoid the extreme mind/body split. We strive to form ideas that are "adequate" in themselves, expressing an independent completeness reminiscent of the *naturans*. Without reference to external objects, these ideas then stake a claim to be sound enough to also reflect God. For ideas are true to the extent that they connect to the God that is the world. This truth is the perfection which *is*: "By reality and perfection I understand the same thing."[8]

Do not think that this is a perfection that is already there as much as our world is real! Like all ideals, *this is a reality which is something to aspire to*. We are not there yet, we have not reached God. Our ideas approach Nature to the extent that they are aware of their interconnections. How does this determined world of ideas relate to the perfect reality? Spinoza asserts that "the order and connection of ideas is *the same* as the order and connection of things."[9] This is a kind of logical form which the workings of the mind are meant to share with the workings of the world. So we discover how the circle evidences perfection, round as it is in our minds or expanding as the ripples extend from a single drop of rain on a glasslike lake. The exact circle is not to be seen, but is perfect because our mind can work like the water upon the wheel.

So how are we to think in ways that approach such perfection, and how should the modes of knowing be cultivated? Here Spinoza distinguishes three levels of possible knowledge, gauged not in terms of relative certainty, but in the different sources to which we ascribe our certainty. The first kind of knowledge embraces two parts: knowledge from the senses and knowledge from signs. In the case of empirical knowledge, sensory experience may directly suggest ideas to us, but in a vague and confused way. But knowledge from signs is a more discrete and definite route to ideas; it explains how the reading of text recollects ideas by specifically pointing to them. It also covers the kind of knowledge that comes through *using* tools, without understanding the principles that allow them to function.

Despite their difference in specificity, both types of knowledge fall under the first kind because they progress purely from individual stimuli, in the first case sensation, in the second case sign recognition. If the knowledge gained remains at an automatic level, Spinoza calls it *opinion* or *imagination*. This is similar to the lower levels of Plato's divided line, with an added emphasis on the perfunctory side of the act of reading, which cannot mirror truth until the material understood evokes the greater principles of the second kind of knowledge.

In this second level, we find knowledge based on our comprehension of "adequate ideas of the properties of things." This is Plato's region of reason*ing,* or knowledge of the principles behind the working of things. Spinoza chooses to emphasize the "adequate" quality of such knowledge: if we understand the principles that guide things, we can predict the behavior of actual objects simply by applying them. Thus they possess a kind of truth independent from the objects to which they may apply. This is a step forward on the progression toward perfection: the ability to predict the logical structure of the world in itself, without external proof.

How does Spinoza's coveted third kind of knowledge differ from Plato's *theoria?* Plato tries to show that mathematics is true because when we finally understand it, we feel a deep conviction of its certainty. Thus it must already exist within us, before our experience of digging it out. This may be tough, but the certainty unveiled is worth the trouble. Spinoza, too, tries to use mathematics as an example of what he calls "intuitive science," but he will not put forth all mathematical truths as example, but only those simple enough so that we may grasp them without effort or intention. It is the kind of knowledge that is peaceful in its certainty, instantly validating a connection between thought and the world at the moment we touch it: "This [third] kind of knowing advances from an adequate idea of the formal essence of certain attributes of God to the adequate knowledge of the essence of things." [10] This is the way to know that allows us to move between the perfection of the one world through the paths of *naturans* to the core of objects. It may greet us easily, but it is far from easy to achieve. Nor is it an explanatory "thing" to guide all else like the Platonic "Good." It is a way all things could be known, if their full connections to each other were to become unquestionably manifest.

For Spinoza, *intuition is an achievement;* the more we know about the world, the more obvious it becomes. In the end we will resonate with perfection, asking nothing from it, accepting the endless implications of each part as much as its limits. We are *implored* to pay attention to everything. This ethics should fall right into place through the linked knowledge of things. Our freedom may stem from Nature, but does it require us to side with Nature? Consider this telling comment on our bond with the world: "We suffer in so far as we are a part of Nature, which part cannot be conceived by itself nor without the other parts." [11] Is the "suffer" here to be wholly independent of unpleasant allusion?

All Spinoza specifically states is that "suffer" means not being in control. So if we are acted upon from without, we may be said to suffer the effect rather than cause it ourselves. We are effected, not causing. Our freedom is diminished because we no longer determine our actions. We do not follow from our own necessity alone.

And this is simply the way we fit into this world; never as wholly free beings. We are always subject to passions and thus still subservient to an order of Nature, adjusting ourselves to the flux of this natural world. This is what Spinoza calls human bondage, again not as a form of punishment but as a simple fact: *we will get nowhere alone, without the world.* This Nature binds us when it makes us passive, frees us when it lets us act. Freedom from our point of view comes from concentration on the active, and from knowing the involving extent of the world in its detail and magnificence.

We need then to work at thinking with the Earth, but not consider its unqualified presence to be anything to fall back upon. Our minds and lives should be oriented toward perfection. This harmonious conception can guide the search even as it is deemed to forever be incomplete. Spinoza resigns himself to the limits of this idea, in a letter to Henry Oldenburg, who asked how we might know the rules of interconnection without ever knowing everything that is connected. This is Spinoza's reply:

> I do not know how the parts are interconnected, and how each part accords with the whole; for to know this it would be necessary to know the whole of Nature and all of its parts. . . . By the connection of the parts, then, I mean nothing else than that the laws, or nature, of one part adapt themselves to the laws, or nature, of another part in such a way as to produce the least possible opposition.[12]

This theme follows all attempts to position technology in relation to nature: how can we assume order in the whole while in the midst of changing it? Some kind of balance is a goal here, a balance in favor of accepting a thing when its context explains it. The collection of relevant data is then less important than knowing the rules, and these rules are in turn much less significant than intuiting the equilibrium of possibilities in a deep but direct manner. Objects known in this way are no longer necessary, possible, or contingent. They simply *are,* required by what surrounds them, then right by the environment. The third kind of knowledge gives them their animate place within the stream of things: "The more we understand individual objects, the more we understand God." [13]

How can Spinoza's austere qualification of ideal knowledge aid in the quest to understand technology? He seems at first to be an odd choice for inspiration; his system has been criticized for being static, oriented towards pure knowledge rather than action. He seeks to explain how humanity ought to fit into the universe, rather than change it. Despite the emphasis on a journey toward intuitive understanding of things in their divine particularity, movement appears to be lacking from the system geometrically outlined in his *Ethics*. Spinoza's system is tested by the suggestion that technology might be included as a vehicle for human fulfillment in the approach of nature.

Change appears in this system in the ways things are related to one another. Nature remains the one substance out of which all other entities are formed, but the myriad of possible strata that may be extracted from it are called "modes," or modifications, of Nature, all those things which may be conceived only in terms of other things;[14] that is, they have no meaning outside their connections to other parts of the total system. Nature is all-inclusive; all other items have meaning just in relation to one another. This is a world of linkage and interdependence, with everything carved out of a single embracing material.

The human mind is this force that may identify the presence of modification while struggling to guess at the way the parts relate to the whole. The Latin *perficere* may express both the movement to perfection as well as accomplishment. Thus action is of paramount importance to the perfect reality which we could strive to achieve. So human ideas of possible ideals are required to assess whether a construction has reached its perfect end or not. Spinoza notes that if we come across a construction site, we may only call the work imperfect if we believe it to be some stage in the process of building a house. If we have no idea what is being constructed, it is impossible to gauge its level of completion. There is no sense of perfection without a model to compare actuality with possibility. And there are no known ideas outside the human mind, so perfection is only possible through humanity assessing its actions as modifications of Nature.

In contradiction to Aristotle, Spinoza announces that there are no final causes in Nature, no goals which natural processes are wont to realize, because the reason Nature acts and the reason Nature exists are equivalent to one another: the same, primary free necessity that makes the one substance encompass all. Final cause, or teleological purpose, is "nothing but human desire."[15] The need for a house is only the final end because someone decided their life would be improved by living in a house. We envision the attributes required to complete this intention, and then the perfectibility of the situation is perceived. Yet as Spinoza quickly points out, "we are conscious of our actions and desires, but ignorant of the causes by which we are determined to desire anything."[16] He outlines an overall human goal of trying to know nature directly, to lead the most active life with the most connections to the most parts of nature. This is an alluring and sweeping vision, though without the specifics necessary to steer actions toward the most active type of understanding. The right kind of technology can show which actions bring Nature closer.

To put any desire into practice, we choose it from among the range of available possibilities. Human action implies a choice among possible transformations of the world. With human freedom, we take a stand on the initial order, narrowing and specifying our concerns. Yet choice is useless without some criteria to help us decide what to do. It is obviously too easy to simply demand those choices which bring us "closer to Nature" or to "follow Nature's harmony." The goal is once more a distant but attainable reality, equivalent with perfection, invoked to temper our desires. When these wants swell with the technical increase of our capacities, the tempering function of nature is more necessary than ever, so that our intentions to transform have some foundation.

Spinoza has outlined a way in which Nature can function both as the ground of all human action and as an end, as we aim to realize the full extent of our connections with the world that surrounds. But the Nature thus reached is an ideal totality, which may be inclusive of change but powerless to explain it. Though traces of Spinoza and Aristotle remain present in all attempts to justify action by nature, neither philosopher was prepared to foresee the great challenges to the very idea of nature which technology would bring.

The following sections will review four major strands of explanation of the way technology modifies the range of the concept of nature. Each will be considered in turn, recognizing the diverse means used to present nature as goal, throughout their activity of redefining and revamping it. All this will lead up to an opportunity to narrow the discourse, testing the quest toward nature upon more complicated technical grounds. A selective historical order is followed to show what ideas make modern technology possible, and then what ideas are used to criticize it. The four thinkers to be considered are: Francis Bacon, who shows how technology makes nature matter to humanity by applying science to the world as resource; Karl Marx, who has technology begin with nature, which is meaningless until transformed into meaning through material history; Martin Heidegger, who has technology precede nature, enabling its possible revelation; and the twin visions of Lewis Mumford and Marshall McLuhan, who unveil driven and searching histories of technology that hope to enable ways of life in line with the potential of nature only humans are able to discover—perfect according to Spinoza, fulfilled according to Aristotle.

BACON TURNS NATURE TO RESOURCE

Spinoza struggled to construct an idea of nature which might encompass the belief that we are capable of improving on the world. But there were probably many more people impatient with the whole grand philosophical hubris that assumes it is possible to *think* anything substantial about something as solid and dirty as nature. In 1531, Juan Luis Vives, tutor at the English court and friend of Thomas More, wrote the following condemnation of abstraction:

> Enraged against nature about whom they know nothing, the
> dialecticians have constructed another for themselves, the nature
> of formalities, of individualities, of relations, of Platonic ideas
> and other monstrosities which cannot be understood even by
> those who have invented them. They attribute a name full of
> dignity to all these things and call them metaphysics. If someone
> has an intelligence which . . . has a bent for abstruse things and
> foolish dreams, they say that such a person possesses a
> metaphysical intelligence.[17]

In contrast, says Vives, the peasants and artisans understand nature better, because they work with it constantly. Nature as a fact in material life leads to quite different conclusions than nature as speculative goal. It is here that we find the basis for those views of nature which stress the predominance of human use over human concept, suggesting an image of humanity as progressing entity, improving through history, harnessing the world's processes unto ourselves.

Not all philosophers of this transitional time were wary of the practical approach. Francis Bacon is generally recognized as the first to formally elucidate technology as *precursor* to science in the advancement of human presence. The very idea of progress is noticed in the successive development of tools long before it is seen in science. Bacon sees improvement in the world around us. He identifies directed change as the organizing quality of nature which we should grasp as a model: "For what is founded on nature grows and increases, while what is founded on opinion varies but increases not." [18] Philosophy demonstrates little progress after thousands of years spent rehashing the same questions, while the mechanical arts advance continually, always showing acceleration. For Bacon, a life spent refining technology is far more "natural" than a lifetime of contemplation. Using the world is the human way to fit into the world. Truth is not only usefulness but power, a means to thrust human purpose outward. Only by applying knowledge do we improve our place in the world. And nature is itself the most fundamental tool.

It is wrong to claim that Bacon *subjugates* truth to application in the technical realization of human meaning. He is rather trying to elevate the status of useful discovery as part of the continued investigation of the riches of nature. Consider the nearly preposterous but immensely valuable resources nature has offered us through our imaginative penchant for discovery. Who would have thought the frightening power of an explosion could be tempered into a manageable tool for human warfare in the form of gunpowder? Or the secretions of a worm woven into silk cloth of surprising strength? Or a magnetically sensitive piece of stone transformed into a device with which to reckon geographic position? These are Bacon's famous examples, hallmarks of achievement that stand as testament to human greatness. It is the fact that they *work* which should be applauded as truth. We are amazed by the success of our inventions after calling them impossible, yet if we allow ourselves to be convinced, we learn a sense of certainty that comes only when the power of a discovery is something that can be controlled:

> The human mind . . . first distrusts and then despises itself: first will not believe that any such thing can be found out; and when it is found out, cannot understand how the world should have missed it for so long. And this very thing may be justly taken as an argument of hope, namely, that there is a great mass of inventions still remaining which . . . through the transferring, comparing, and applying of those already known . . . may be deduced and brought to light.[19]

Here is how the countenance of the mind is affected by the accelerating pace of new invention. First we learn to trust in the boundless potential of the future. Then, working with current techniques, we gain confidence that improvement will follow naturally. For technology brings meaning to discovery through invention, finding new things in nature, fulfilling them by putting them to human use.

Bacon considers technology to be a model for the future of human culture, as it shows how it is possible to press forward. And philosophers, with their dreams of ideal states and worlds, have always wished for improvement. Activity as opposed to meditation allows people to work together, and it is only in the refinement of devices and instruments that *perfectibility* has actually been observed. The culture that embraces movement will be a technical culture, one not afraid to shape the surrounding world towards its own stated designs.

There can be no progress if one believes that *techne* merely completes nature, rounding out a natural perfection through a realized human sense of place. For Bacon, it is *techne* that permits us to participate in nature, to be a part of creation and movement on equal footing. Making and transforming is our single entry into the natural realm of progress. He challenges Aristotle by stating that

> the artificial does not differ from the natural in form or essence, but only in the efficient. . . . Nor matters it, provided things are put in the way to produce an effect, whether it be done by human means or otherwise.[20]

Techne is the addition of humanity to nature. Because we can build things as nature builds them, we are entitled to call ourselves part of nature. This might seem today to be a most shameless attempt to justify anything humans may do to the world with the name of nature, but at the time it was a theoretical revolution designed to help people gain confidence in the value of their own abilities. Technology brought tangible proof that the achievements of Renaissance civilization had eclipsed the classical ideal it was claiming to reinstate. Philosophy in its ethereal purity continued to look regressively backward; only practical knowledge held the prowess to manifest the future.

What would this future look like? The utopia of Bacon's *New Atlantis* is not so much a paragon of lawfulness and cooperation as it is an ever-burgeoning state built upon the principles of human extension. The Order established in New Atlantis to discover how to make use of nature is named *Salomon's House,* where wisdom is gauged in practice. Here what may be sensed in nature is carefully harnessed and transformed into the new, miraculous but always controlled. Behind this ideal litany of research and development sounds a moral and aesthetic tone of the natural:

> The end of our Foundation is the knowledge of Causes, and secret motions of things; and the enlarging of the bounds of Human Empire, to the effecting of all things possible. . . . We have furnaces of great diversities, and that keep great diversity of heats. . . . We have also perspective-houses, where we make demonstrations of all lights and radiations. . . . We have also sound-houses, where we practise and demonstrate all sounds and their generations. . . . We have also perfume-houses; wherewith we join also practises of taste. . . . We have engine houses, where we imitate and practise to make swifter motions than any you have. . . . We imitate also the flights of birds; we have some degrees of flying in the air, we have ships and boats for going under water, and brooking of seas; also swimming-girdles and supporters. We have divers curious clocks, and other motions of return. We imitate also motions of living creatures, by images of men, beasts, birds, fishes, and serpents. We have also a great number of other various motions, strange for equality, fineness, and subtilty. . . .
>
> And surely you will easily believe that we that have so many things *truly natural* which induce admiration, could in a world of particulars deceive the senses, if we would disguise these things and labour to make them seem more miraculous. But we do hate all impostures and lies: insomuch as we have severely forbidden it to all our fellows, under pain of ignominy and fines, that they do not shew any natural work or thing, adorned, or swelling; but only *pure as it is,* and without all affectation of strangeness. These are (my son) the riches of Salomon's House.[21]

With progress in the mechanical imitation of nature *should* come a vow of truthfulness. Nature achieves meaning to the extent that it may improve humanity, and the society based upon such a belief devotes the majority of its energies to the expansion of human faculties. Inventors and discoverers are lauded with monuments and rewards; prayers are said not only to God for "his marvelous works," but for the illumination of "*our* labours, and the turning of them into good and holy *uses.*"[22] (But look who is not lauded: the meek, the powerless, the cheesemakers.)

It is not a vision far off from our world of today, populated by a humanity eager to praise the wondrous order of the natural world, while also quick to pat itself on the back for being able to discover these riches and put them to humane use. Bacon's dismissal of the fear of tampering with the natural order is essential before one may believe that human position in the universe may be heightened by the result of our actions. Anyone who believes that technology can change things for the better will want to follow him this far. The danger looms when nature is said to be meaningful *only* to the extent that it helps humanity. Human purpose becomes the one purpose that matters, justifying a wholesale appropriation of everything we may discover about nature to make the world better for us. Anything we find that does not support such improvement is best ignored. Technology becomes the sole criteria of truth.

This is the standard interpretation of Bacon, usually backed up by a passage in the *New Organon* where truth and utility are called the same thing. But Paolo Rossi argues convincingly that a more accurate rendering of Bacon's Latin supports the following translation:

> Things as they really are, considered not from the viewpoint of appearance but from that of existence, not in relation to humanity but in relation to the universe, offer *conjointly* truth and utility.[23]

This amounts to stating that the world as nature includes human use, and is neither opposed to it nor subsumed by it. Human comfort is certainly not enough of an intention to drive technological development; rather, utility is philosophically important because it offers a road to the truth within things. Using the world, we have a chance to recognize its inherent value as well. Bacon encourages us to accept that knowing comes through making; he faults the Greeks not for their belief in absolute truth, but for their elevation of contemplation over action. Bacon sets the stage for a truth that will be the destination of human discovery, detectable only through devices we can construct after gleaning fragments of nature's own processes.

Truth cannot be the same as utility because we do not know everything, yet we are able to build and make things that extend our knowledge. Any invention involves hypothesis, risk, and testing. We envision an instrument that may aid us, but we try it out without knowing if it will work. It is possible for our plans to be faulty, so the human drive to improve must allow for mistakes. There is no linear completion of a fixed destiny. Some tools may work better than others, making utility relative. Truth, however, holds its firm place as the overall goal, reachable by welcoming technology, not spurning it.

Bacon's move to unlock the secrets of nature, liberating them from the zone of the forbidden, is a prerequisite for the installation of applied science as a pillar of human development. Because placing faith in technology changes the way we conceive the world, it is also quite precarious. It is easy to conceive of utility as truth, much harder to comprehend what it would mean for both these qualities to be attributes of things in themselves. Yet this is the challenge all of us must answer as long as we want to believe that humanity changes the world and still remains a part of nature. If we alter the world, truth itself is affected. But we need not make the mistake of replacing it by our use of nature.

Bacon restates the puzzle, framing our bond to the world in a way that teaches us not to be afraid of our desires to build, discover, and topple the scaffolding of familiar dogmas, constructing solid tools of progress in their place. Yet the maxim of "Improvement!" is not enough of a prescription to save and direct humanity in nature; techniques are still more valuable as "pledges of truth" than as contributions to the spread of comfort in our lives. The difference is that improvement can clearly be applied to specific situations, once we have accepted its call. Truth remains a global notion, still abstract and distant, however strong its claim on us may be. The search for technological advances may proceed smoothly upon the realization of exact human intentions, without these intents being governed by the guide of the true.

Spinoza's netlike conception of humanity willed by perfection toward the world came only a generation after Bacon put forth his *New Organon*. It is possible to accept the latter's rules for human conduct as a means for the approach of the overall intuitive fit advised by the former. Indeed, grasping the precise place of things in the world order will be direct and intuitive only if we know them not in detached, passive contemplation, but in the kind of active acquaintance that comes through working with things in practical life. Bacon never wants us to forsake nature, only to chisel ourselves more deeply into its riches by imploring us to use the world.

Did he worry that we might use the world up? I think not—the ideology of his time was so dominated by fear of worldly or divine wrath that he felt people needed more shaking up and encouragement than any restraint with regard to the world's resources. But embedded within Bacon's radicalism lies respect for the goals of the Ancients, if not their methods. There is a truth of nature, and it is up to us to seek after it and be led by it. The problem is that particular progress is so much easier than looking far off for advice. Once we apply our efforts to technology, it is amazing how fast we can upgrade our abilities once our needs are meticulously defined. The need for truth is not like that. As much as progress comes so easily to the machine, advancement in the realization of truth fades far into the horizon.

We start out with Bacon wishing for truth and utility within things, as we end shaping truth *out of* utility. Any truth nature might possess becomes impossible to prove apart from our use of the world. Bacon's call to arms was far more successful than he could have imagined, while his call for knowledge recedes as we become more amused by our toys. Today we possess and understand many of the inventions prophesied in *New Atlantis,* but without the systematic happiness they were supposed to bring. This may be because we forget that accomplishment synonymous with perfectibility requires a standard like "nature" in order to be judged. Otherwise we have only empty motion at the center of the system: individual improvements may be charted, but without any overall intention to drive the movement of humanity.

It is tough to retain a sense of forward current in the progression of human history if we hold onto a fixed idea of nature that threatens to curb us. How can the future remain an open and unwritten world if there is an order that wants to hold us in? Is it that much easier to make than to understand? Future generations saw less need for Bacon's cautionary plea for a planned world that includes our transformations, instead making due with his encouragement to extract what was needed from the world, changing this to imply "go beyond the world!" Nature as a guide became far less convenient in the explanation of humanity— too much was happening far too fast.

Whether he saw it or not, Bacon's decree for technological pathbreaking challenged the very possibility of a truth apart from utility. The idea of progressive usefulness is powerful and convincing. Nature itself could not "work" the way new devices could work unless it also might embody such swift progress. The more we value betterment of the human condition, the less we can accept the place where we began as an absolute value.

The natural world was changed into mere beginning, necessary only to start human history rolling. We transform nature into meaning through working with it, and this meaning is historically developed to realize human community, now entirely mediated by the tools we develop to make ourselves out of the world. This view does not deny importance to nature, but accepts it only as material, no longer as the far away perfect order which we occasionally get a glimpse of. Technology remains the force that propels the change of meaning through history, but nature is now only the matter that makes progress possible. This ambitious view is fleshed out in the work of Karl Marx.

MARX IN SO MANY MIRRORS

Take away the fluxing energy of the *naturans* of Spinoza's Nature and we are left with *naturata*, material, nature as world of extractable objects. This is the place where Karl Marx imagines history to begin, as pure substance out of which humanity is able to forge meaning through *praxis,* making objects out of the world which form our own nature external to ourselves. The fact that our essence lies outside us, out there to be made, shows we are a part and parcel of nature:

> Humanity needs objects as objects of its life-*expression.*
> A natural being has its life outside itself.
> A sentient being is a suffering being.
> The true human life becomes the *externalized* life.[24]

With these enigmatic fragments, Marx hints at the way humanity expresses itself by constructing meaning with material things. As natural beings we are dependent on the environment in many of the same ways other animals are, while at the same time endowed with a need to objectify the world. This life in the midst of objects is not without its price. Living in a world of things we have made, we become detached from our essence in the drive to ensure the world's presence. Seeking meaning, we are driven away from meaning.

Marx precisely articulates the technological danger that Bacon was unable to see. He attacks the abstraction from practical being in the direction of *either* humanity or nature—their truths are inextricably bound to each other through work and understanding. For Marx, *technology is primary being,* because that is where humanity becomes itself through projection into the material world.

Yermiyahu Yovel suggests that Marx continually struggled with the limitations of language in the expression of this fundamental location for human being, fighting ceaselessly with the pair "humanity and nature," so that their dialectical synthesis would offer a practical sense of the right way for us to live.[25] Marx was trying to express the notion of this polarity as a unity, a "human-in-nature" expressing the primal extent of our species.

Marx sees *techne* as the most material of human activities and thus the one from which all else follows. By beginning here, he does not dismiss nature or human essence, but shows their dependence on practical life. Then is technique alone enough to sustain the progression towards better and better human societies? An answer requires comparison of the Marxian idea of the expression of human essence through work with the idea of technology as human extension.

Technology is the first fact of humanity for Marx. Living through *techne*, we transform nature into history and thus into meaning. We cannot do this just by thinking about our place in the world, but only through working with the world, digging our fingers into the soil, cutting down the trees we need to build houses and rafts, reshaping, moving, and carving a world out of nature. Because we are human we look for our world outside ourselves, inside the realm of constructed objects. Externalization places ourselves into the objects we make. Through practical life, we leave our human living limits and enter into artificial tangibility. If we are able to find something of ourselves in the things we create, they are not completely detached from us, but somehow express us. It is the tendency of humanity to want to express this essence by externalizing it, that is, by making and believing in objects distinct and separate from us.

This rhetoric of Marx may seem cold and "technical" to some, yet it is also passionate and convincing. He claims to begin with reality, in the thick of the lives most people lead, examining the things most people construct and believe within the grip of praxis. Technology is our conduit to nature, with its ever advancing promise of the better way of life. The problem is that in the drive to externalize our lives, we leave something behind. The object, however fulfilling its use may be, remains unable to express humanity in full splendor. We make it, we want to realize ourselves through it, but it stands there exuding emptiness and disappointment. This is alienation, the darkest quality Marx attributes to the material base of human existence:

> The worker relates to the product of his labour as to an alien object. . . . The worker puts his life into the object and this means it [the life] no longer belongs to him but to the object. So the greater the activity, the more the worker is without an object. What the product of his labour is, that he is not. So the greater the product the less he is himself.[26]

Two questions immediately come to mind after this account of the receding of our creations from our essence and desires. First, if the fundamental activity of humanity is technological life, just what is alienated through the producing and using of objects? Second, is alienation present in all kinds of materially guided life, or only in those in which workers are exploited—denied access to the fruits of their labors and forced to toil mercilessly for the lazy bourgeoisie in a society out of balance?

Neither is answered unequivocally by Marx himself. Alienation seems to be present in objectification itself, and there is something about this latent discontent that drives humanity to want to build more, and to progress upon the pull of dialectical opposites seeking impossible reconciliation through the perfectibility of human history. We are also told that some societies are better than others, either because they have learned how to temper inevitable alienation, or because they have learned how to produce things "in a human manner." But what is this attribute "human"? How can it be conceived apart from the technological struggle to make and be materialized in the external world? If alienation is a danger inescapable by human technique, then there should be a humanity preceding technique, or at least in place to be *saved* from it. But such spiritual or metaphysical notions will not be tolerated within dialectical materialism.

Technology does not simply alienate the natural from the human. It is not encompassing nature that seems foreign to us, only a nature distorted through humanization.[27] Our tendency to imprint ourselves onto what we extract from nature turns against us through those political systems which encourage bigger and better objects, externalizing with unforeseen vigor and dimension. Marx does see a light beyond the angst at the end of the tunnel, which will be visible only in a revolutionary social structure where the destiny of our products can be controlled wholly by our connected selves as both producers and consumers.

For alienation itself quickly moves from the existential to the political. Making results in having, and the private or restricted ownership of the things we create for each other alienates us again, this time from the inherently social aspect of human nature, in which property is shared. Technology threatens humane society as much as it ensures its durability. Giving ourselves up to things, we lose our rapport with people. The more complicated the things are, the harder they are to produce, and the more people will lose themselves in the process. The more artificial and constructed the world becomes, the less any construed nature seems to matter in our drive to externalize ourselves. This may be why the natural world has so limited a place in the utopian consequences of Marx's vision. But if humanity is to continue to live in a world that is more than our own externalizations, nature must also be expressible. External nature need not spurn all things human if the things in which we trust so much are made and exchanged in a different way. The social constraint of capital requires inequality. Instead, we need to come up with a way to value one another in terms of practical, mutually recognized worth, independent of so technical a result as monetary exchange value.

Marx would doubtless find the concept of technology extending individual intent far too isolationist to express the way our made world helps us. He would say to me: "Of course in your eyes your product is an instrument, a means to be able to control my product and thus to satisfy your needs. But in my eyes it is the aim of our exchange."[28] The creating individual obsessed with results, seeking some solace in the object of their labors, is bound to be disappointed. If I instead consider my work to be successful only to the extent that it participates in mutual exchange with others, then I will be able to find a place in society and not be destroyed by an alienation born of the innate desire to build *things*.

Marx asks for *production in a human manner*. In describing its qualities, he speaks from the first person conditional: how would I know when I have reached it, if it were socially possible? "I would have objectified in my production my individuality and personality." So these are the parts of humanity threatened by alienation! Need they be protected by technology, or must they be drawn out by it? Communication is impossible without tools of transmission and abstraction, so technique should not be conceived as contrary to the expression of the individual. Marx is critical of an isolationism that wants to define the individual as distinct from all others rather than as cooperatively social. The humanly producing individual becomes himself by giving from within to others, not losing social place by identifying too heavily with possessions. We are saved only if we may appreciate how we are appreciated by others: "I would have had the direct enjoyment of realizing I had satisfied a human need by *you enjoying* my work of the human essence."

Technology now serves not to externalize while depleting the internal, but to reinforce inner humanity by holding people together. And a bond from one person to another soon leads to identification with the whole human project: "For you, I would have been a mediator between yourself and the species." You give something from within *through* the external object *to* me, planting the seed of social meaning in the human community. Our course in the world is enhanced through constructive dependence on one another. Expressing ourselves may draw us in toward others, and need not always end in detachment: "In my expression of my life I would have fashioned your expression of your life . . . thus realizing *my communal essence*."

This is Marx at his most mystical, and his closest pass at any notion of an inner human truth. A human nature does appear, and it is a place where the objects of our labors do not hide us from ourselves. In this world view every tiny gesture is directed towards reception by others. From the cognizance of any single human creation, the meaning of humanity emerges if we truly comprehend how the device is to be used: "In that case our products would be *like so many mirrors*, out of which our essence shone."

How like Spinoza this sounds, when the lens maker was asked to explain how we might envision the whole if all we may be certain of are the relations of some of its parts! Each segment of the world reflects the whole of the world to the extent that we know the ways it connects to all other parts.[29] So the more we appreciate the influence of our works, the more *techne* builds confidence in our sense of human place.

But for Marx it remains all hypothetical, framed within the speculation of "what if we *could* produce in the human way?" He does not believe we can as yet attain this goal because he fixes his attention on objects made rather than tools that allow one to make—only a detached result can provide proof of the alienability of labor. The dehumanizing effect of repetitive and meaningless work is vindicated most graphically if the process builds things of no use to the worker. What is useful to the laborer would bind one person to another. Only self-representation of the social character of humanity does not alienate.

Despite its analytic function within the dialectical explanation of historical change, there is something frozen about this vision of workers confronting the fruits of their labors, downtrodden with the inability to find themselves therein. It pits the producer against the product, locked into a binary, either-or situation with nothing tangible in between. The fault lies in the mechanism of the Hegelian thesis-antithesis formulation, whatever synthesis in social revision is prophesied. However dynamic it purports to be, there is something static in the account of the worker making objects while seeking subjects. The materialism invoked by Marx as a call to reality denies the fluidity that allows us to create in the first place.

The plea to produce objects that express social reciprocity ignores the fact that technology changes our conception of what society is and can be. Breaking the technical process into the maker and the made betrays the original humanity-nature continuum that Marx claimed to be primary. He is unable to flesh out this promising beginning because he singles out only one cumulative effect of technique on history as deserving of attention: increasing exploitation of the many for the profit of the few, with these groups severed from each other by the unequal social arrangement enforced by dependence on capital.

Certainly the conditions of his time justified this observation, and no civilization could have survived the industrial revolution without a slew of drastic changes. But Marx did not see the flexible nature of technology itself! Suppose it is not just an expression of humanity which is sharpened in opposition to humanity—we losing ourselves as the machine finds itself. If technology is instead an extension of human intent, our move with it into the external world does not require a break with humanity, or any betrayal of inner contentment. With technology as extension, the designed object carries no meaning apart from human purpose, pushing humanity farther outward in the world.

Alienation is only present if the instrument is a failure, if it subverts our intentions. The pitfalls spotted by Marx become indicators of technologies that are losing the link to their human motivators, starting to stare back at us instead of furthering us. But technology does not grind to a halt like this: remember the flip side of the circle of extension. Technological development opens up far too many new alternatives than we know what to do with. Revealing choices, it does not tell us how to choose. Perhaps this is another kind of alienating stagnation: paralysis in the wake of too many possibilities without criteria to judge them. We feel lost because our machines cannot advise us on what to do with what we have made.

There may be confusion, but a world spurred on by technical change does not let ideas sit still. Marx could not envision how technology would alter the reality of the industrial system enough so that machinery would be exploited more so than men, women, and children. Automation would take many of the more repetitive factory tasks out of the hands of people and into mechanisms of greater intricacy. Management would grow into a vast *logos* of hierarchies and layers trained to make only the most precise of decisions. The industrial system breeds a society of specialists, where each of us finds meaning in the community only when we can perform the very specific task assigned to us.

As machines become more pervasive, repetitive work is taken away from us. We workers must then learn what is necessary to keep the machines moving. Spending more time in the mechanical world, we learn to think in ways congenial to the machines. Eventually repetitive *thought* processes are taken over by tangible extensions of the regularity observed in our minds: there is not much but machines to talk about. We further just those ambitions which can be understood as information, digested into bytes, stored on digital disks, transmitted effortlessly and electronically over wires and airwaves across the globe.

Today a group of workers can collaborate and exchange products without ever seeing or meeting one another, combining ideas from one side of a continent with those of the other. What kind of community will come out of this? Whatever we attempt to agree on, consensus is only possible if prefigured by technology, which has made the felt unity of global human society possible at last. We have seen the Whole Earth from far off in space, and we can pick up the phone and call anywhere on the planet for the price of a single work by Karl Marx. What would it mean to produce in a *human* manner today?

Marx knew that we have no humanity apart from technology, but he, as bound by history as any of us, could not see how deep technology could work its way into the roots of human ground and limit. True, too many workers still toil under the conditions of hardship he detested. And political oppression continues under just about any good name in government: democracy, communism, socialism, or anarchism. But the world has changed more and faster in ways perpendicular to those that Marx decoded for us. Technology has dug more than a single furrow linking humanity with nature. Responsible for the order we have increasingly imposed on the world, it supports this world like a frame which can be glimpsed in itself, isolated from those parts of thought and sense which it is still unable to capture.

HEIDEGGER FRAMES THE EARTH'S PICTURE

In his earlier writings, Martin Heidegger recognized the difference between a world simply present and a world that announces itself as potentially useful, examining the being expressed by tools in themselves. The clocks and street lamps described in *Being and Time* and discussed here in chapter 1 reveal purpose in nature: the moving circle of time, the cycle of light and darkness, and the human need to keep track of the former so that we may modify the latter. Saying that these devices comment on nature means accepting a separation of the technical from the natural. Later in his life, however, Heidegger decided that the concept of technology is much bigger than the way instruments announce their function—so big, in fact, that it circumscribes all our contemporary attempts to live and to know.

The concept of equipment or *Zeug* is expanded into a form that prefigures all our understanding, practical or speculative. Technology is now said to *Enframe* (*Ge-stell*) all of modern life. The power of today's technology is so pervasive because it adopts a fundamentally different way of relating to nature, such that it is impossible to consider nature outside of the bounds of this technology. Is total technological envelopment of the world to be the culmination of human development? No, it is merely another stage we must pass through. It is a preparation for the possibility of human fulfillment that has always been with us: the revealing or saving of the Earth, as Heidegger puts it, "set free into its own presencing."[30]

The universal expansion of technology which so foiled the historical progression of humanity envisioned by Karl Marx may be a boon to Heidegger's dream. *Techne* in its original sense encompasses all human making, deeply tied to *poiesis*, a composition ("setting-into-work") of truth which evokes nature without claiming to explain it. The road to a human life which learns at last to let the Earth be what it is meant to be needs to pass through the period where technique prefigures all. This is a most dangerous time, as the poetic grasp of our place in the world is put into peril by the hubris we instill in ourselves by *using* more and more of the world.

For Heidegger, technology is neither instrument of freedom nor alien object, but an approach to truth. Instrumentality unveils aspects of nature. It has always done so: the windmill reveals that the wind may turn wheels to grind grain, the shovel discloses the riches inherent in the soil below the surface of the ground. Tools show us what the world may offer, previously always in direct interaction between ourselves and a specific natural feature. With modern technology, the interaction changes. Today we may *extract* energy from nature, and store it for later use. Nature is no longer equivalent to the energy we need to survive, but the *source* of this energy. Energy gains meaning apart from nature, when it is removed and held away for future human transformation. Now we look to the natural world, and consider at once what we may take out from it to further our purposes. We no longer accept the river as a course of water in the landscape, but instead start to calculate the number of gigawatts we can harness from it through hydropower plants. We gaze at the Earth, and call it all resource.

This is the eventual result of the pragmatic, forward-looking gaze that Bacon inspired us to follow. Yet he was not ready to use it to end the special human dream of truth. Heidegger is equally unwilling to forsake this perennial philosophic goal, but he recognizes that the increase in our ability to take what we want out of the world makes nature into a source of riches which only human ingenuity may see and *remove*:

> The energy concealed in nature is unlocked, what is unlocked
> is transformed, what is transformed is stored up, what is stored
> up is, in turn, distributed, and what is distributed is switched
> about ever anew.[31]

This is a description of those technologies which are material and embodied, in terms of human extension, instead of in contrast to nature. We now shape our material world out of things taken from the natural world, like electricity, or even things abstracted from it, such as information. These are the entities which are stored, moved, sent, received, grouped and regrouped after the initial step of extraction.

The world waits to be taken. It is no longer the environing context of our aspirations, but a reserve, standing by for our call. Heidegger sees the Earth *as if on stand-by*, poised for our next move, our next renovation. As entire resource, it is ready-for-us. It is as if modern technology were a lens that allows us to see the world as only one special set of offerings and possibilities, completely invisible without the distorting filter of revelation as instrument. We frame the world within the shape of these techniques, and we perceive the world as fuel for human opportunity.

Heidegger does not tell us how technology made the step from directly revealing the forces of nature to transforming what those forces mean by placing them within the new framework of storage and retrieval. Was it a sudden insight, or a gradual chain of discoveries, that led to Enframing? This is not discussed. Since Enframing is so comprehensive, it is impossible to conceive of its limit from within its walls. Technology embodies its own purpose and cannot then see meaning outside its own needs. Seduced into purpose through technique, we are unable to conceive of any vision of the world apart from potential material for our manipulations. Any sense of detached contemplation or wonder fades into the background.

Among the ways of knowing which recede is *poiesis*, which Heidegger identifies with that part of human creativity which allows "what is present to come forth."[32] It speaks of an openness toward the world which is at the same time not afraid to make and to build. This attitude is blocked by the Enframing: we have lost "the shining-forth and holding-sway of truth."[33] A nature that lies behind a locked door, ready to be opened only by the key of our uses, will not be available for discovery.

This is the edge of how technology changes us, denying us the chance to see anything outside its powerful defenses. It is so successful that it eliminates the uncertain hold required by less verifiable modes of apprehending the world. Heidegger is careful in his polemic against the Enframing essence of technology: it is not human making as such that is so dangerous, but one way of construction which endangers all others because it works so well unto itself. Though this "menace" thrives and increases, it is at the same time a necessary rite of human passage. This is an era of danger which we must weather and survive. It is technology that allows us to endure as we are. Enframing charges *techne* with a *logos* characterized by a "frenzy of ordering."[34] The less explicable human glimpses of truth beyond use are trampled by the absolute certainty of design and decision. Yet Enframing does keep us going, changing and exploring. It keeps humanity part of the flow characteristic of nature. And we are needed in this flow to let nature appear.

In his own wafting description of its essence, Heidegger is really neither for nor against technology. How could a blanket acceptance or condemnation of the essential basis of our society be of any use? Because of the strange way he uses words at the ambiguous edges of arguments, he often appears to be straddling the fence where one expects him to take sides. Which life is better, the old way or the new? That is not really the point. Heidegger sketches a route for us to get far enough from the frames of technology to grasp its pervasive essence. He then outlines the attributes our alteration of the world would need to possess if it were to *release the Earth unto itself.*

It is through language that nature is revealed to us, and it is through language that nature has been transformed into material. (Marx would say it is through work.) There is no language without humanity, so nature has no meaning until we are here, making and shaping with words and actions. Yet we are never in complete control of the power of our words. (Spinoza would say that we never know the whole even when we speak of the whole.) Heidegger says that language speaks the Earth through humanity.[35] This defines our largest purpose. We use language only by participating in it. Thus it differs from Enframing, which cloaks and rearranges the world before being able to talk about it. But language is also *techne,* as we abstract through it to create a world that we may communicate to others, as demonstrated in chapter 2. The world-building of language contrasts with the world-denying of extractive technology. Within the former is a clue for the ills of the latter.

Language brings truth forth with *poiesis,* and it is this kind of making and building which holds Heidegger's hints for how *techne* may be rejuvenated. The power of the poetic lies in its ability to measure and to sing the world into existence, announcing the place and rhythm of humanity within the whole. There is poetry not only within words, but in ways of changing the world by building our lives within it. Does this ask for too much from art? It is an attempt to recover a sense of making the world which at once encompasses technique, artifice, beauty, and construction. Forging humanity and the revelation of nature in one and the same project, poetry is the essential quality that *releases* the rest of reality unto itself:

> This measure-taking is itself an authentic measure-taking, no
> mere gauging with ready-made measuring rods for the making
> of maps. Nor is poetry building in the sense of raising and fitting
> buildings. But poetry, as the essential gauging of the dimension
> of dwelling, is the primal form of building. Poetry first of all ad-
> mits human dwelling into its very nature, its presencing being.[36]

We dwell in that we build. Poetry measures and constructs, ensuring the possibility for authentic human dwelling. But this *poiesis* is nothing so definite as the lyric arrangement of language according to rhythm and image. Playing on the inexplicability inherent in all creation true enough to earn the distinction of "poetic," Heidegger tries to use the word to evoke whatever qualities need to be possessed by that technology which will save us and let the Earth pass through us.

Those of us caught in the thick of the struggle to develop real solutions to immediate problems might find the ephemeral answer of poetry quite far from the direct difficulties we face. The poetic seems almost the antithesis of the technical, being so ineffable and rare, elusive to the diagram or the flowchart aimed at results. But because poetry appears so far away from technology, it may be able to break through Enframing and guide the reformulation of technique.

Poetry is not reflection on the world for Heidegger, nor the milling of words out of context to evoke sentiments inexpressible any other way. Poetry shapes our world, such that when a poem works we feel its truth beneath all subsequent thought on the subject in question. Here is yet another way to suggest the intuitive certainty dreamt by Spinoza. It is another metaphor for the moment when we grasp the essential correctness of a particular action or way of conceiving part of the whole. Whether you accept or reject it depends on your impression of the magic of language beyond a rational ability to denote and signify. If you can glimpse truths in language that shine like crystals behind all the arbitrariness of individual words involved, something of the genuine human place in the universe comes through the abstraction of letters and sounds—out come the mysteries which make possible *life*.

Humanity has the potential to free itself from Enframing because we are able to "dwell poetically" on Earth. The poetic is the name Heidegger chooses to give to that elusive quality present in all things that we or the world may do to allow truth to come forth without forcing it. (In chapter 6 others will use other names for the same elusive idea.) A technology that forces resources out of nature will never be poetic, but an instrumentality open to natural powers may well be able to wax poetical. The vision is appealing, but diffusive. It suggests a complete about-face from the techniques of today, yet as a goal it should be accessible from the present situation if it is to be realistic.

We can find poetic transformations among the many changes technology unleashes on the world. Windmills and sailboats do connect us to the powers of air without removing anything from the sky. They do help us fit into the world in ways road building does not. But nature changes as we learn new ways of knowing it, making the natural hard to hold onto as an ultimate aspiration.

Heidegger tries to solve this dilemma by harkening back to the dynamic, growing senses of *physis*. As the primary realm of building, "*physis* is indeed *poiesis* in the highest sense."[37] Nature is the only entity entitled to continuously construct itself, endlessly opening and expanding without needing to cross boundaries. All other efforts are technical, and thus partial ones: they take a piece of dynamic nature, and become dynamic themselves only by producing something outside themselves. A human creation will never in isolation or in specific use be pure poetry. But if it is correctly made and used, it may partake in the grand *poiesis* of nature.

The vision wafts closer to the mystical. *Techne* at its best is meant to encompass *poiesis* and *physis*. In its original Greek sense, *techne* as the sum of all arts was meant to be singular yet manifold, keeping truth secure by allowing it to shine through all doing, acting, and human changing. Techno-logy, on the other hand, becomes something else when the initially supportive power of order carries the predictability of action too far from the ineffable nearness that is nature. It then surrounds us with a rigid grid of our own making, with each part a precise knot in the net of human purpose. We cannot cut our way out, even as we behold a unity that makes life appear more balanced before.

What is the appeal of the continued call to act more *like* nature? Why does it continue to surface above all charges of irrationalism and weak sentimentality? It is easy to understand the human impulse to hope that the way we live collectively as a species is *right by the order of the world.* Civilization aspires to meaning and stability when it justifies itself by nature. This is the old Aristotelian dream, explicating and shaping a vision of the individual and society which makes sense according to rules greater than human ingenuity or inquisitiveness alone. The quest of science (in its widest sense) is to diagram the absolute connections which bind humanity into the universe. If we live in a way that enforces rather than weakens these links, we survive by recognizing our intended part in nature.

The power of such a view is affirming and expansive. By living in the correct manner, we associate ourselves favorably with empirical forces still beyond our control and reach. We participate in an overall order beyond our comprehension simply by allying our understanding with it. Taking the side of nature could never be a mistake, as long as nature stands for an amalgam of those tangible but irreducible agents which extend before and after humanity, holding a universe around us. This is the only kind of *poiesis* which exceeds us, yet no one will recognize it unless we call attention to it through our own history of works. That is our special place in the scheme of things.

The biggest problem with this argument, held throughout history by many besides Heidegger, is that as human society evolves, the qualities supposedly possessed by nature are continuously redefined. Sometimes it is the wild and savage enemy, and other times it is our solace and comfort. Do we approach closest to it while contemplating the wild, untouched forces of wilderness bounded away from civilization, or in the thick of tilling the land, gently shaping it to yield fruits we may eat?

Each age answers the beckoning call of nature in its own way. If this answer includes the reshaping of the Earth through technology, the nature of this planet is in itself transformed. To believe there is some authentic Earth that may shine through all our manipulations is part of the vision of an interaction with nature beyond the fragmentation of technology. But poetry still builds its wholes out of parts, and words themselves come out of the tool kit of language. The moment we claim to know something of nature, it begins to be humanized. If there is an inhuman essence to nature, we will never grasp it, since the ways we reach for it are part of those aspects of human *techne* which are always reconfiguring themselves. Still the beguiling aesthetic of nature remains present in so many attempts to chart the course of the future, even as technology becomes omnipresent across the globe. Although we cannot reach it, we need to hope it is there.

THE MYTH AND THE MESSAGE

Meanwhile, nature as a unified, though distant, inspiration still exerts its pull on us. The qualities we ascribe to it may always change, while a constant and recognizable attraction still entices. In this section, two twentieth-century thinkers on technology are compared, both of whom invoked the "natural" as an attribute of those technologies which marked the correct future for humanity to follow. Lewis Mumford called the achievements of mechanical and social engineering "organic" in the middle of the 1930s, hoping they would transform the exploitative growth of the industrial period into advances that could be enjoyed by all individuals, harmoniously pushing forth toward a renovated sense of human life. Marshall McLuhan coined the current use of the word "media" in the 1960s to name the advances in communication likely to link all persons on the planet together so that we may directly reach one another, as one gigantic tribe, in a global village accessible to all. When Mumford reaches this later time of upheaval, he condemns McLuhan's superficiality: if we surrender ourselves to blind faith in the transformative power of new technology, we will forget what nature might mean beyond ourselves. The machine becomes a myth, no longer of promise engaged with the environment, but of salvation upon irrational, empty hope.

For Mumford, the apprehension of our condition and its movement should never begin by *deference* to context. Every life form asserts itself, taking a stand:

> Every form of life is marked not merely by adjustment to the environment, but by *insurgence* against the environment: it is both creature and creator, both the victim of fate and the master of destiny. . . . In humanity this insurgence reaches its apex, and

manifests itself most completely, perhaps, in the arts, where dream and actuality ... are fused together in the dynamic act of *expression* and in the resultant body that is expressed. As beings with a social heritage, humans belong to a world that includes the past and the future, in which we can by our selective efforts create passages and ends not derived from the immediate situation, and alter the blind direction of the senseless forces that surround us. To recognize these facts is perhaps the first step in dealing rationally with the machine.[38]

From the outset of his investigation into the insurgence of technology in history, Mumford wants to judge all successive technical development according to the criterion of *life*. Life proposes biological ends, and each movement in technology must be assessed in terms of the faithfulness of its effort to realize these ends. With this in mind, he puts forth a conception of three great phases in technology, the *eotechnic,* the *paleotechnic,* and the *neotechnic,* with the first extending roughly to 1750, the second running from then through the industrial revolution until the turn of our century, and the third gathering momentum in the 1920s and 1930s, extrapolating promise into the generations that were to follow.

The *eotechnic* is marked by handicraft, agriculture, and the direct embodiment of human extension into the environment through our own labors. By expanding human presence outward, this phase enhances human life by harnessing the immanent natural forces that surround us. Water, wood, and stone are the dominant materials. Tools are manufactured by craftspeople for specific tasks, easily customizable by the user because they are simple enough not to require mass standardization. It is for Mumford a generally positive time, with technology enriching the life of the senses not only through direct perceptual extensions like the telescope, but through developments in urban and garden design as well as the artistic depiction of daily existence. The surrounding world expresses humanity in the landscape, from the symmetrical patterns of fortresses to the cultivation of fine wines. Back then we expressed ourselves in nature's terms.

Why did we move on from it? The change began in England, somewhat at the fringes of the eotechnic establishment. The population increased dramatically, and the old agricultural order could not be sustained, at about the same time that it became economically feasible to consume energy out of the Earth in the form of coal. The new industry was not based on life enhancement, but on the extraction of a source of power out of the ground itself, fueling transformative inventions like the steam engine and the railroad. Human presence is expanded into nature, not directly, but through machines driven by coal robbed from the landscape. Life is quantified and driven, run by the intertwined materials of iron and carbon. Workers had to suffer to make these industries grow, and the physical environment was sacrificed as well. It was capital that drove this advance into squalor. This was the period which Marx labored to explain. But for Mumford (and anyone else who lived into the twentieth century), this period was not the culmination of human innovation, but a preliminary phase in which human strength is tested against nature, only to learn in the end that life must be upheld once more. This is why he calls the industrial age the *paleotechnic* era, a passing phase in which quality of human life was sacrificed to further the prowess of technology. To transcend it, we should step back from its unique brutality to affirm a higher humanity in a living world.

This is more or less the point where Heidegger leaves off. As far as he is concerned, the technology of the twentieth century shows no signs of deviating from the devastating directions of the nineteenth. But American pre-World War II optimism renewed faith in technology, looking once more to nature, adhering to a new organicism in line with that preached by Henri Bergson. Human culture would find a way to progress and at the same time follow the guide of life, evolving toward a new technically inspired unity with nature.

Mumford names this time of promise the *neotechnic* era, when technology fulfills its original purpose by bringing humanity and the expanded world back together once more. This is the time of electricity, of social engineering, efficiency, and the birth of instantaneous communication. This is the period when the machine begins to arc back and affect human essence in disarming ways. Cooperative thought, the functionalist aesthetic, and a more balanced, material sense of human personality are some of the effects of mechanization upon the mind which Mumford applauds. Such new influences can be for the better if we recognize that, "in projecting *one* side of the human personality into the concrete forms of the machine, we have created an independent environment that has reacted upon every *other* side of the personality."[39] Comprehension of the modern machine makes order accessible to all, no longer the sole privilege of an industrial complex ruled from one place above.

Though this order is to be a democratized nature, Mumford is not naive enough to claim it as an absolute, separate from our progression towards renewal of humanity with the aid of the machine. No, even in his steadfast belief that the machine may be inducted into the service of life, nature is never to be wholly independent of human inquiry:

> We may arbitrarily define nature as that part of our experience which is neutral to our desires and interests: but we . . . have been formed by nature and inescapably are part of the system of nature. Once we have picked and chosen from this realm, as we do in science, the result is a work of art—*our* art: certainly it is no longer in a state of nature.[40]

Mumford encourages us to assimilate the mechanical virtues of impersonality, objectivity, and neutrality *before* we can sail towards the edge of the "more richly organic, more profoundly human"[41] civilization that returns to the virtues of life. So in a sense he is a confirmed mechanist, while maintaining faith that mechanical thinking will show the way beyond itself.

What evidence does he give us to support the conclusion that technology has radically changed enough to see past its limits? Machinery itself is no longer composed of standardized, identical units which are pieced together. Instead, individual parts are specialized, refined, made more precise so that they may fit into the whole and achieve meaning in the great technical system. Earlier machinery needed to simplify organic processes to render them in mechanical form, while neotechnic mechanical parts become more complicated individually the better to mirror the compex niches of the natural world. The game of machines is no longer like checkers, with teams of identical pieces, but like chess, with complexities of strategy. The goal: dissolving the rigid mechanical world picture and redirecting us towards an organic understanding, quantifiable only in terms of growth and change.

The shift toward the natural in technology begins when we come to see machinery not as the towering achievement of an ingenious humanity, but merely as "lame counterfeits of living organisms."[42] What is an airplane next to an eagle, a radio next to the voice? Our proudest technical achievements only approximate the organic functionalism within nature, and by reconsidering the wonder of these natural processes, human techniques can be rejuvenated.

Mumford's examples tend to stress the "natural" aesthetic success of modern technical thought: a set of ball bearings, a spiral piece of a metal spring. Do these things not evidence a basic beauty? (He notes that they're in New York's Museum of Modern Art.) The more pure and abstract these objects are, the farther removed from their context, the easier it is to appreciate them as manifestations of pure form, composed of circles and spirals, the roots of a pristine, mathematical nature. They are like Plato's Ideas in action, pure cast and polished shapes, at last put to work in the practical world. No wonder they suggest nature as exact and certain, finally manifest in real, useful objects! As human production specializes, we imagine we can build and think more like nature, and may stand more chance of approaching the magic of natural processes with our refined and meticulous minds.

But it is with Mumford's illustrations of supposedly "organic" architecture "harmonizing" humanity and nature that his optimistic and reforming technical vision begins to appear suspect from the present vantage, sixty years later. A photograph of boxlike, concrete single-room dwellings for Swedish workers framed against an evergreen forest is touted as a "handsome and well-integrated human environment, in which the efficiency of neotechnical production can be registered in a higher standard of living and a wider use of leisure."[43] Seen from today, it looks like a row of mobile homes parked at the edge of the woods, ruining the view. Not to mention the consequences of uniformity upon those who live there! Right beneath is a photograph of a hydropower station, as rectilinear and devoid of affinity with environing nature as any concrete skyscraper glimpsed across a prairie. For Mumford it is a "symbol of a fresh mode of thinking and feeling." It looks like a piece of empty, colorless industrial America today, taking no account of any sense of place.

Why do these images appear woefully artificial to us now, if they seemed to herald the clean, purposeful lines of nature in the 1930s? We are no longer able to see these structures as surprises, since they and the grids that guide them have expanded so rapidly and easily that the wild has been trampled in their tracks. Two generations have gone by, a Second World War, and advances in technology which threaten to overrun the Earth by their tremendous *triumph*. It is impossible to look at the mainstream of modernist design and affirm that it has approached the directions of nature, *unless* nature now is thought to include powerful forces both harmonious and massively destructive. Then it is not much of a goal anymore, but a fact of the times. (Chapter 5 will discuss how the potentially devastating limits of technology challenge the conception of our bond to nature.)

Yet *nature remains alive as an alternative,* home to a more profoundly human life always beyond the horizon of present capacities. The pull was there for Aristotle, Spinoza, and Mumford, and it remains with us today. Aspiring to nature is striving for a way of life which will be settled into the world around us. At times it has meant taking control of that world, and at other times it has implied fitting in. The invocation of nature as a goal for human progress has been a mix of insurgence and deference, to use Mumford's terms. As the encompassing goal that both upholds us and is malleable to our desires, nature is the most immanent of possible human ends. That is why Spinoza wished it to be God. And it is chosen by Mumford as the one measure to which technology needs favorably to compare if it is to be judged appropriate to the forward gaze of this century. It brings our ideals near at hand.

With this perennial appeal of nature in mind, it is all the more striking that Mumford's concrete box examples work no wonders for us today. The world heralded by these innovations has shown little sign of approaching the organic qualities so revered in the twenties and thirties. Of course, this kind of building was well-intentioned, yet it is not enough to say that its adherents were not aware enough of its tendency to outstrip its surroundings. While the normative power of nature stays with us, with its share of dynamic and balancing qualities that admit human progress but benignly check it, the specific cultural moment requires that the link to nature *never* be held back from change.

This would make nature not a freely relative term, open to whatever attributes we choose to ascribe to it, but an ideal that takes new forms whenever a society decides to illustrate the components of a better world to be achieved in the foreseeable future. The disturbing upshot of this is that the analogies to nature might only be those that agree with the conventions that mark that particular instant in the evolution of society. The farther we side with such a claim, the more empty nature becomes as a possible goal. Could this be the dreaded end called postmodernism?

Nonetheless, Mumford's pleas for a more organic technology seem as appropriate and as unanswered now as then, while only the examples he puts forth seem misguided. This lends some support to the idea that the guiding concept of nature is best left abstract and unsubstantiated, the better to be applicable from one era to another. Does it then become too vague to mean anything? It never goes away—we will always dream of a way our ever-expanding order can fit into a pattern greater than that which we may conceive. Using the word *nature* brings this concept down to Earth. It encourages attention to previously unrecognized connections between things and activities. We learn our relations to the world only if the world is first conceived as a scheme of relations.

For Mumford this sharpens a caution just beneath the surface in his early work, which is brought to the fore in his later: that *technics will never encompass all human activity*. While agreeing with Heidegger in stressing the importance of the conversion of energy as the turning point behind modern postindustrial technology, he dissents by still locating primary human meaning beyond technology, not within it. He looks for the freedom technology offers us as result, not as enhancing process:

> The real significance of the machine . . . lies in the gains of energy through increased conversion, through efficient production, through balanced consumption, and through socialized creation. . . . A society in which people worked to live and lived to work would remain socially inefficient. . . . If we are to achieve a purposive and cultivated use of the enormous energies now happily at our disposal, we must examine in detail the processes that lead up to the final state of leisure, *free activity, creation*.[44]

The machine makes possible an easier life with more opportunity for human diversion. This world will be technical to the extent that it becomes "efficient" (the buzzword of Mumford's era). He is right to prophesy a society in which leisure will burgeon in economic and political importance, as greater quantities of our time will need to be filled, hopefully in ways more creative than deadening. And in this extra time, technologies rise to amuse us in every free moment we get.

Marshall McLuhan follows Mumford's lead here, calling attention to those techniques which expand into the vacuum of leisure time. He christens them *media* and finds therein encouraging signs of a society fulfilling every optimist's dream: taking hold of the best out of the golden past, while adapting it to an ever rosier future. A generation after Mumford's plea for organicism, and a generation before the present commentary, McLuhan sees how instantaneous electric communication radically alters what it means to speak, write, touch, feel, and live in an expanded human community rendered possible by technology. This world thrives not just on the storage and transference of energy, but on the holding of information as well. What is most promising about such machinery? Once again, it brings us back to a kind of "natural essence," a neo-tribal village that spans the virtual globe.

What is the harbinger of all the good news? Electricity. Initially it was a source of consternation and uncertainty, an unfamiliar force for humanity to reckon with. By the 1960s it is taken for granted. We have even become bored with it and are letting it sap our own strength. This, says McLuhan, is because we have failed to let the technology seep enough into our thoughts. In other words, we have not let the technical result transform our intentions.

Electricity shows us two faces. On the one hand, it is devoid of content. This does not mean it is a mere means for the realization of ends that require energy. No, the electric light, for example, should be recognized to be *pure information*.[45] It is either on or off, with no options in between. The meaning of such a technology is not the specific opportunities it creates, helping us find our way in the dark. The message of a medium is something larger: it is the entire corpus of changes introduced into human life by the new technique. For McLuhan, too, the change is embodied outward—technologies achieve purpose when they extend humanity towards the world. The meaning of electric light, then, would begin with the extension of the activities of our lives into places and times where natural light will not follow us.

But that is not the whole story. Electricity is more than the indicator of information. It has a moving, fluid aspect, pulsating through the artificial fabric of linked-up humanity connected to the global power network. Electricity plugs the various parts of the world to the same actual wire; then it synthesizes the analytic, divisive thought of the digital West with the inhabitation of process associated with the East. "The contained, the distinct, the separate—our Western legacy—are being replaced by the flowing, the unified, the fused."[46] McLuhan compares electricity with the ineffable Tao of Lao Tzu: nameless, empty, but essential in its flow. The whole Earth comes together at last upon a principle that cannot be caught.

These are his two classes of media, perceptible in all communicating technologies which involve abstraction: *hot* and *cool,* bridging the gap from one generation's hip jargon to the next. Hot media extend one single sense to the exclusion of all others, filling our perception of that one sense with a great profusion of data. Photography and printed language are given as examples, as they overdetermine image and word respectively. Cool media present a field of meager information, leaving much for the perceiver to fill in. Hieroglyphics serve as a mnemonic for spoken language, and the voice filtered through the telephone always leaves much of the person speaking to the imagination of the one who listens. These cool media encourage participation of the receiving parties, while the glut of precisely directed information burning through hot media leaves us in the cold, as we have nothing to add.

The distinction caught on so fast because it is so simple, judgmental, and nearly impossible to pin down. One person's fire is another's ice: the newspaper may seem an endless reel of facts to be thrown away after an hour's digestion, while the same words can be cut up into poetry and recited very slowly. McLuhan probably realized the pliability inherent in his distinctions, with hot and cool properties latent in any medium, alternately coming to the surface as the medium is refined and applied. When he vouches for the superiority of the cool, he is not claiming to make a momentary judgment of style. He wants to describe those journeys which may be made by way of media to advance civilization towards its most significant goal. That destination is once more a place in nature, and here McLuhan's message resonates with what his predecessors have been saying.

Then the cool medium is more natural because *it requires more* from all those who are connected by it. It inspires more, because it gives us only vague suggestions that achieve results just with our imagination and involvement. Nature becomes once more the domain of a community of equals, at one with the world only when each of us must give something to the network to be enhanced by available information. This is an updated vision of Marx's community of mutual producers, only now it is not as important to make objects as it is to make personal sense out of the plethora of concurrent facts racing past us at any given instant.

The fragmentation we sense around us is a residue of mechanical thinking, which solves problems only once the elements of a situation have been broken down into discrete components. The precision parts of industrial machinery grind against each other in their performance, with each part functioning in one set way in the service of the whole. What electricity adds to this system of interaction is a flow of energy independent of the work accomplished. It runs through wires, linking individual parts instantaneously. They may now perform several functions, and participate in a network more reliable than the scrape of one gear against another. Interconnected and instant knowledge is now available across the grid of power. There is no mechanical limit to the speed of electric transmission. Electronic constructions may now approach the spontaneity of life. This is the root of the analogy: "It is that same speed that constitutes 'organic unity' and ends the mechanical age that had gone into high gear with Gutenberg."[47]

What Heidegger cautioned us against is what McLuhan wildly applauds: *automation* is the word he invokes to name those technologies which work indirectly upon the stockpile of energy. Machines that can function merely by plugging a power cord into a wall socket promise tremendous freedom, like a living organism defining itself through an ecosystem. Faith in this comparison is what leads some to believe that the omnipresence of technology today is itself enough to "save the Earth." As each part of the world learns to touch the rest of the world on equal footing, unification of the planet becomes inevitable, as long as we continue to conceive the process as the approach to nature:

> This same need for organic interrelation, brought in by the electric speed of synchronization, now requires us to perform, industry-by-industry, and country-by-country, exactly the same organic inter-relating that was first effected in the individual automated unit. Electric speed requires organic structuring of the global economy quite as much as early mechanization by print and by road led to the acceptance of national unity. . . . On the other hand, the old-fashioned kind of "war" becomes *as impracticable as playing hopscotch with bulldozers*. Organic interdependence means that disruption of any part of the organism can prove fatal to the whole.[48]

The very presence of electricity is a clarion call for us to expand our notion of community to enclose the whole. The more we link ourselves together with the tether of electricity and information, the more the human system which spans the whole Earth resembles an organism, the manifestation of nature, the image of the correct way to live. McLuhan even has enough faith in globally assimilating technology to imagine that it will prevent the practice of war!

This picture of correctness has its roots in the ideal of the village, where face-to-face human contact is the norm, and each person participates in the community by constantly playing diverse and loosely defined roles. The small-scale organization possesses an inherent stability through the diversity of tasks performed by each person. This is the oft-imagined model life of the past, when each of us would have been able to wear many hats and never be threatened by a pigeonholed existence.

Yet as villages succeed they tend to grow into cities; as their walls become fortified and concrete, they become like skins to their inhabitants—the limits of an extended world. Territory is divided and conquered. There is no more room for the flexible richness of village life. The rise in mechanism encourages compartmentalization, until electricity opens up worldwide instant communication between equal individuals, at the dials of their telephones. As the world in its vastness becomes directly apprehendable through the media of today, intimate and profound contact with a tremendous range of people, cultures, and ideas becomes the hallmark of the age. If McLuhan is right, this swift expansion involves an equally remarkable sense of return, realizing the ideal of the village at the vast limits of human scale:

> Ours is a brand-new world of allatonceness. "Time" has ceased, "space" has vanished. We now live in a *global* village . . . a simultaneous happening. We are back in acoustic space. We have begun again to structure the primordial feeling, the tribal emotions from which a few centuries of literacy divorced us.[49]

These are strong words. They suggest a full-scale historical cycle, in which the best of the past may be recovered simply by putting the right foot forward. By accepting cool technologies as instruments to coax participation out of us, we may constitute a complete human community out of virtual materials and invisible networks. The key is choosing the right tools, or perhaps more accurately, finding the right side of our tools—the side that invites us into new encounters by remaining ambiguous, as opposed to the other side that overburdens us with a fountain of information and precision beyond our ability to assimilate.

It is strange to think back on the introduction of this distinction as part of the "birth of the cool," heralding a decade in which the ideals imagined to lurk in the future were thrust back in the moment, demanded all-at-once by a generation in rebellion against the ruling authority. McLuhan's part in this new kind of global revolution was to sanction the search for pieces of the new dream within the technology produced by the very system whose upheaval was the goal of protest. The way to do this was to insist that one of the most invasive of advanced communication machines, the television, draws out of us a hidden potential that might help recreate a tribal human idyll that is anything but "nasty, brutish, and short," rather, the exemplary form of *living* in a human manner.

As is often the case, the attempt to take an aspect of a single technology to be the symbol of a cultural potency neglects the inevitability of technical change. McLuhan calls the TV cool because it offers fuzzy, black and white images on a tiny screen, from which we must greatly extrapolate if we are to participate in the story or event depicted. The indeterminacy of the televised picture is its strength, as we must engage the machine with our imaginations if it is to actively extend our reach and perception. The vagueness of the medium makes it a conduit for expressing humanity, but we need to give a lot of ourselves to identify with the paltry, flickering picture on the imperfect tube.

It is hard to maintain today that TV is a subtle, suggestive device that spurs the creative mind to complete its open-ended reel of stimulating, visual ideas. Broadcast and reception technology have improved in tandem, and television is now more crisp, in color, and in stereo. Wide screens approaching the size of dwindling cinemas are now affordable for large living rooms. Over a hundred channels are available through the more deluxe cable services, so there is more for the viewer to choose from than ever before. Programs are spliced together with remote control; we watch ten at a time, fleeing from commercials. Future developments promise improved picture definition that will bring the home screen closer to the film and the photograph, offering precise information in both visual and aural poles, warming up the medium so that little trace of the cool remains. And yet the quality of the programming has improved only sporadically. (With special interest channels it is "hotter" than ever.) The medium should never be the only message.

Now the only TV set I own is a black and white 1960s vintage model, bought from a philosophy professor at a tag sale for two dollars. It gets just a few channels, and each of these comes in with a surreal, perhaps organic inconsistency, complete with a veil of mystical patterns which may cloak the program, rendering it barely comprehensible most of the time. To pay attention to any program of major length does require much extrapolation. It is a more demanding experience than watching the same show on an up-to-date, full featured instrument. If I want color, I imagine it. If I want to empathize with the action, I can only suppose it to be larger than it appears.

Taking this primitive tool to be television, it *is* possible to imagine how McLuhan came up with his concept of the video box as invitation to join in with a world of immediate, though inadequate images. But the technology has moved in a direction perpendicular to the involving, communal, and idealized natural tendencies he saw in it. Technology will not stop with a situation that encourages ambiguity. The reasoning that makes technical innovation possible is wont to improve the impreciseness of any present moment. If the televised image is in any way underdetermined, somebody will be working to fix it in the warmed-over name of advancement. There will be more, clearer, and sharper options to choose from. Technology is never content to leave alternatives to the user's imagination, as long as what is left out may be conceived to be a technical problem.

So McLuhan's assessment of the promise of TV becomes a momentary side effect in the actual evolution of visual broadcast technology. He has noticed that there is something intriguing about the incompleteness of this pioneering phase. But just as this is spotted and lauded, the progressive nature of engineering has gone on to correct it. Technology has always been invoked to warm the cool, to ensure certainty and reduce risk. The tool itself cannot deal with what it has not been designed to encounter, even if its users can. Yet technological change tends toward systems that operate smoothly. If we cannot precisely explain the organic *flow* between human and machine, the connection remains outside the technological project. New technology does not teach how to improve the content of information which courses through the airwaves, unless this content can be phrased in technical form.

This becomes a warning to all attempts to plot the meaning of a given technology: do not propose a purpose that denies the tendency of technical reasoning to correct its quirks as mistakes! With this in mind, consider the tremendous furor McLuhan caused at the first publication of his ideas, and compare it to his lasting effects. At the time, reviewers felt they needed to have an opinion on the most radical thesis of *Understanding Media,* which was that writing as a medium might be totally eclipsed by the more palpable and immediate neo-tribal conduits of television, radio, and the then hypothetical personal computer. The public outcry was tremendous on both sides.[50] Today, despite growing computer literacy and the proliferation of hypertext, writing continues to survive, perhaps because in its singular linguistic overdetermination, it allows for many kinds of ambiguity. As technologies shift and multiply around them, *words* still hold onto a firm piece of ground. Maybe they are cooler than McLuhan cared to admit. Words and their enveloping system of language have proven flexible and malleable within the most stringent of media, and though we change the way we use them, their visual appearance in print, even if on a screen, remains a part of most hyper- or multimedia experiments. If we venture to call them a technology, they are certainly one of the most enduring.

Concurrently, many of the changes heralded by McLuhan as part of the new worldwide tribalism are often touted as detrimental today: fragmentation, overstimulation, overimmediacy of information with no one medium exerting any primacy over any other. We have seemingly endless choices at the supermarket, the stereo, or the video store, without any accepted criteria by which to judge the cornucopia of opportunity offered by contemporary society. Can this be our nature? Today we are more apt to call "natural" those directions which provide universal checkpoints for meaning, which encourage the establishment of an overall context for humanity, beyond the extension of opportunity. The world appears endangered; what is natural is its preservation. Today's dream is to ground humanity, to learn to consider the Earth as a home again. Will we be able to flesh out this vision so that it becomes more than a temporary response to the current state of changing technology? Remember, there are no grounds to expect that technology will learn to respect those problems which elude quantification.

McLuhan was right to note the shortening of our attention spans in response to the growing barrage of information which we constantly confront. But this led too often to a kind of blind acceptance of the coincidental, the inevitable synchronicities of a world in which too much is going on. In French the word *mcluhanisme* became the label for artistic happenings in which diverse forms of expression were combined without the shackles of any overriding order which might hinder our enjoyment of the chance manifold of aesthetic opportunities of one medium poised upon another.[51] As inhabitants of the world village of media, we should already possess the habitual skills necessary to tune into such an event. Art of this nature has received a certain amount of acclaim and acceptance as part of postmodernism, but it seems to leave something to be desired for most of the audience/participants. What has been forsaken is any claim to be responsible for the outcome of the work, and commitment to principles is nothing that media can substitute for by using the name of nature.[52]

In his willingness to wholeheartedly embrace media, to wait and see what happens, McLuhan seems in the end to avoid the difficult task of directing their inevitable evolution and transformation. He completely evades the location of responsibility for technology, and neglects the fact that authorities may direct the potential of technology to empower or to repress in directions that are more conducive than disruptive to social control. This is a point made by Hans Magnus Enzensberger in a theory of media outlined several years after McLuhan's big splash.[53] The German poet and cultural critic notes that every radio, telephone, or television receiver is a potential transmitter, suggesting an inherent emancipatory side to media. However, in practice, there is most commonly one transmitter and a whole slew of receivers, making the medium primarily an instrument for the broadcast of centrally controlled programs, rather than an instrument of networked, decentralized communication. He calls this a repressive use of media, because it reinforces the immutability of the current system by spreading its tentacles in the form of messages that never stop and that may never be altered by the listeners and viewers who try to shut them off.

"Personal" computers and modems may appear to change the situation somewhat, but first it should be recognized that very few people across the globe have access to the technological infrastructure necessary to support such devices. Second, most who use these devices are not in full command of them, instead being at the mercy of software developers always in the business of trying to sell new products that promise more open and egalitarian means of communication between one machine and the next. In no practical sense is this worldwide web of information "free" and constitutive of an encompassing community. It makes full communication the property of only those who are part of the elite world of the most advanced technology.

To some extent these advances do trickle down to all who are in the business of spreading information of any kind around the world. There are more ways to send and receive data instantaneously than ever before. But whether this constitutes direct and natural exchange is another matter. A tremendous and reified political system is necessary for information flow, and the presence of this flow does not replace the need to temper technology by accepting the blame for it as residents and perpetrators of the system. Humanity should not give up any sense of innate responsibility through faith in the autonomy of technology, because such faith forgets the origin, function, and context of techniques as human extension. If we wish to sail toward a nature, it had best be a *human* nature before it is one of grand and austere design.

This is the conclusion reached by Mumford in the magnum opus of his later life, the two volume study collectively entitled *The Myth of the Machine*. Thirty years after *Technics and Civilization,* historical events had shown no sign of turning technology toward the organic ideal. There had been a second, more devastating war that introduced a greater potential for human-induced destruction than anything previously conceivable. Technological enthusiasm led to tremendous postwar growth across America, with roads, highways, and settlements entirely determined by the techniques available to create them quickly, efficiently, and homogeneously. None of this, says Mumford, is working. Technology has failed to come around to nature. The popular voices of the 1960s are preaching anarchy. No one is willing to take responsibility for the mistakes that have affected all of us.

This is a deeply pessimistic vision, revealing a bitterness of the man betrayed by the history he had hoped his ideas would have altered. He recognizes that he may have been too frivolous in his youth, placing too much weight on technics as the prime mover of civilizations.

Wholesale belief in the saving power of machines has led to contemporary crisis: we can now be saved only by ourselves. We need to accept that humanity is always *more* than the sum total of our accomplishments and inventions. Technology, according to the later Mumford, not only transforms intentions but *forgets* them, caught up in the quest for streamlining and efficiency. In a scathing self-criticism of his own earlier work, Mumford chastises his own tendency in *Technics and Civilization* to attribute the positive qualities of human intentions to the machines themselves, "qualities that often *disappeared* at the very moment the technical processes themselves were being simplified and perfected."[54] Here the machine grows into a myth, appearing to promise values it is unable to assimilate or embody, as ethics remain solely a human province. So it is up to us to take hold of the spreading, phenomenal success of mechanism, which meanwhile threatens to seduce us into complacency.

Mumford's example of the worst kind of complacency is the anarchic world suggested by McLuhan. This global dissolution of culture into a "tribal communism" devoid of literacy is anything but natural. Mumford is most incensed by the call to accept randomness, to sit back and soak in sensations while lauding this aimless wandering as the culture of the future:

> In McLuhan's trancelike vaticinations . . . even the wheel is
> about to disappear, while humankind as a whole will return
> to the pre-primitive level, sharing mindless sensations and pre-
> linguistic communion. In the electronic phantasmagoria that he
> conjures up, not only will old-fashioned machines be perma-
> nently outmoded but *nature itself will be replaced*: the sole
> vestige of the multifarious world of concrete forms and ordered
> experience will be the sounds and "tactile" images on the con-
> stantly present television screen or such abstract derivative in-
> formation as can be transferred to the computer.[55]

No tribe is united by the omnipresence of audiovisual media. As disembodied streams of sensation, they cannot constitute human community. "Real communication is only possible between people who share a common culture—and speak the same language."[56] One cannot suddenly invent a new language and expect it to establish interpersonal understanding. The fallacy is that information is not enough to guide life. Human culture involves practices and emotions which cannot be subsumed by the onslaught of data.

On the other hand, the globe does feel much smaller today than ever before. We are all subject to an increasingly assimilated blend of aural and visual stimuli. Teenagers in tiny villages in the mountains of Nepal are as likely to give travelers their opinion of the latest pop record as they are to give directions to Katmandu. Is this enough common ground to suggest that the world is one culture? Mumford says no, and the biggest mistake is to equate sameness in image with the human diversity that is *our* nature. This dream of equal access is the opposite of the organic, which only admits choices that increase a sense of belonging to our communities: "What is needed is a technology so varied, so many-sided, so flexible, so responsive to human need, that it can serve every valid human purpose. *The only true multi-medium remains the human organism itself.*"[57]

Herein lies the paradox of technology as route to nature. Why call it "technology" if what is meant is simply the total realization of human purpose? Mumford wants technology to be like the human organism. And yet he also recognizes technology to be an extension of the human organism. From the outset technology changes humanity, so it is destined to be unlike that part of humanity which cannot be extended. On the other hand, if it is technology that *defines* humanity—if humans are the only beings who can totally transform their environments to serve their purposes—what notion of nature could be compatible with human purpose, except one wholly malleable to the mercy of our desires?

The dilemma suggests the impending death of the idea of nature as any kind of overall guide to humanity in the throes of unstoppable motion. Even Mumford, the stalwart champion of the organic, has turned the idea on its head in specifying the *human* organism as the final model, as if our knowledge of even this closest of organisms is independent of those tools which immediately extend its bounds. Now the instrumentality definitive of techniques is no longer considered to be enough to enable our achievements. In *Technics and Civilization,* smooth, rounded precision machine parts were pictured as evidence for modern civilization as a society that could manufacture and apply the perfection of Plato's geometrical forms in practical life. Now in *The Myth of the Machine,* there appears a stone figure that looks like a model of a submarine screw propellor, but is in actuality an ancient Peruvian archaeological find, with no known practical application in that society.[58] The message: innovation of geometrical forms is another example of a human idea that *precedes* any real-world usage (in this case by a few thousand years). Idea comes before action, intention before realization.

This is Mumford transformed by history into a rationalist, with technology more consequence than inspiration.

Strangely enough, the more recent book does not then make technology seem less important by placing activity after theory. Instead it increases in significance, largely because of the negative effects that it recounts like a litany of errors committed in the name of unmanaged progress. Although twentieth-century history has revealed the failure of the earlier wish that technology might right itself, Mumford is by no means ready to give up the organic analogy. On the contrary, he thinks it was not introduced broadly enough in the earlier work. Nature is more than the external temper of technics. It must be conceived so that human aspirations and nonextendable faculties may be included as well. A nature that is only the context of humanity may have ended, but a nature that includes humanity needs to begin.

Need we fault Mumford for still believing that a "biotechnics" will come to eclipse "megatechnics"? The biggest question remains whether there is any unambiguous notion of *life* apart from our technical attempts to assess, some *élan vital* which is unified in our inability to catch it. McLuhan disturbed Mumford enough to inspire the latter to see the need to carefully distinguish the diversity of life possibilities from the overabundance of choice delivered by a consumer society built on the diversity of specialized machinery:

> Under the power complex the purely quantitative concept of unlimited abundance . . . has served as the guiding principle. As opposed to this, an organic system directs itself to qualitative richness, amplitude, and spaciousness, free from quantitative pressure and crowding, since self-regulation, self-correction, and self-propulsion are as much an integral property of organisms as nutrition, reproduction, growth and repair. Balance, wholeness, completeness, continuous interplay between the inner and the outer . . . aspects of existence are identifying characteristics of the organic model.[59]

An organically conceived and managed world would offer us not only a plethora of choices, but an assembly of only the *right* choices, which build upon one another, offering meaningful options that do not dissipate, but organize and identify. For Mumford, the variety of life forms and processes in the organic world is not random, but ordered as each part influences the other parts, like scaffolding to support the whole. These kinds of diverse choices regulate and motivate themselves, sustaining the motion of life forward as they constantly change.

This combination of sustainability and dynamic mutuality is contrasted to the push towards more and more choice and variety characteristic of our present culture, which lacks integrative meaning or necessity. We may buy any of one hundred brands of cereal in the supermarket. Are any of them really any good? Does their diversity still hold up beneath their packaging? Each frosted bran flake tastes the same. Advertising is employed to make us want one before another. This diversity is concocted to impress us, making the choice more difficult but not more profound. Life as a consumer is made more complicated, but not more meaningful, with colorful ways of dressing up the processed and the efficient so that it may appear interesting.

A Norwegian disciple of Mumford and Marx, Sigmund Kvaløy, distinguishes this kind of *complication* from an organic and natural *complexity,* in which diversity is only present when meaningful in terms of the whole, with the many clearly connected in explanation of the one.[60] Every blueberry off a bush tastes slightly different from any other. No snowflake resembles another, yet their diversity reflects the endless creativity of a nature that works freely upon rules. Human society may imitate this with the creation and usage of objects that reveal the individuality and the special place of whoever made them. So grandma's secret recipe is never in any book, and when you eat it you enjoy not only the food but also all your memories. Human lives are not standardized, and we assert our direct personality through things we make and use ourselves. This is a renovation of Marx's "production in a human manner," resuscitated to stand in sharp contrast to the tremendous range of standardized choices available to us today. Organic plenitude would admit only those variations which could be shown to make sense in terms of each other, in which we do not lose ourselves, but instead find our place in the world as a whole.

Yet technology continues to fight this blissful vision of natural integration, where humanity gradually casts its net into the whole, aiming to discover exactly where we are and should be in terms of all things around us. Why does the turn of the technical toward the natural always seem *just beyond* the horizon, a sensible goal that we continue to veer away from at the moment it begins to seem attainable? Because technology *works*; it succeeds at the realization of particular human intentions. The moment it is complete, possibilities for improvement are also suggested. Alternatives appear. These become technical alternatives, *not always options we would have wanted to consider in the first place,* but variations that demand to be tried out.

As technology becomes increasingly enmeshed in its own rigor and hierarchies, the reasoning of efficiency tightens its grip. It is far more difficult to speak of anything alive once seduced by the functional language of the machine. Even Mumford is lost, as his utopian organicism sounds like just another manageable product, "regulating, correcting, propelling." His hoped-for future still lies within the discourse of machinery. Technology changes nature into something it is able to talk about. Thus subdued, the natural is powerless as any kind of imperative to set a course by. Yet we continue to wish it is there, without knowing how to see it or to speak it. Denying its existence forgets the fact that the ideal of nature does not go away, remaining an eternal, if ephemeral force vying for an exemplary place in our lives. The myth remains a message, even as the tools we construct to view it conspire to transform the content of our gaze.

ARTIFICE DIVERTS NATURE IN TIME

This chapter has charted a rough course from nature to humanity and back through the troubled and transforming means of technology. From Aristotle to the present day, nature has exerted its appeal as a goal for human knowledge and then progress, while its precise meaning has been impossible to pin down. As humanity has gained more confidence in the promise of technology, nature itself has appeared to us in increasingly technical terms. It becomes harder to conceive of it as any kind of goal anymore.

For Aristotle, all things that exist by nature are propelled towards realization of their natural potential, approaching an ideal sense of belonging, the location of happiness. With Spinoza, humanity is a singular part of the one eternal substance because we are subject to emotion and free will. It then becomes our task to work at approaching nature not in the way we have come out of it, but with new technique: We need to know according to nature before we can live according to nature, and this necessitates a very particular quest for certainty. Perfection becomes the knowledge of how one relates to everything else around and within, not listed or calculated, but immediately felt. This model frames a vision of humanity circling through nature, which later authors challenge.

For Bacon, perfection starts by recognizing perfectibility: we may alter our situation. We make our place by transforming the environment, fearless of risks, never sure of the consequences. Several hundred years later, Marx assesses these rampant transformations and calls for a reconsideration of the value of production. Making things is the human way of being natural, if the right kind of community is constituted by our actions, one in which each of us knows and fulfills our rightful place.

Heidegger discloses that we have lost the essential natural side of our penchant to make and to build. This is poetry, which nature embodies to the fullest degree. But modern technology in its isolation and storage of energy is deaf to the call of interaction with the world. Because it is so diametrically opposed to *poiesis,* Enframing sets the perfect stage for us to strive for its negation, so that nature can shine through once more. Mumford also sees an organic turn as the culmination of modern technology, with the best technic being one that wants to be natural. McLuhan witnesses the fulfillment of Mumford's dream, seeing an instantaneous natural world society, linked up like the firing synapses of a global brain. The present, empty moment is the river of Heraclitus, the *logos* in which we do and do not step. Mumford tells him he is wrong, calling his vision the most pathetic substitute possible for nature, as it reduces interconnection to sensation after sensation, empty of meaning.

If one were simply to ask these thinkers to list the attributes of that technology which would be worthy of the name *natural,* they might answer in the following ways. Aristotle: it helps us fit into the world, completing what nonhuman nature has left unfinished. Spinoza: it comprises those tools which enable us to know the world as God with direct, unmediated intuition. Bacon: it perfects our place in the grand scheme of things by making possible concrete action that leads to physical improvement of our situation. Marx: it would be those methods of production which allow us to link with and need one another, rather than alienate. Heidegger: it would reveal why nature needs us, by permitting us to dwell poetically. Mumford: it would work like nature, sustaining us by enhancing our connection to the organic principle behind all living things. McLuhan: it would link us all, instantaneously, as nodes on the great network of life.

Common to all these views is a hope that technology will help us find a home in the face of forces we will never completely understand. Yet technology concurrently alters our conception of these forces. How would each consider technology to alter the idea of nature? Aristotle and Spinoza would say not at all, as nature remains God, or the overall principle of rightness—monolithic and unchallenged, even by human freedom. Bacon: truth will *appear* as utility, thus nature will seem like material, and we will assess its value in terms of what we may make from it. Marx: nature achieves meaning as it becomes history, which culminates in a human community as dynamic yet sustainable as nature alone can be. Heidegger: the Earth becomes what lies beyond Enframing, once we learn how to set it free. Mumford: we start to see nature in refined, pure, and exactly machined objects that assert a beauty equivalent to function. Later nature comes clearer in opposition to the insensitive, contextless mechanisms that now fill our empty spaces. We then resuscitate life by idealizing it as an alternative to the currently visible madness. McLuhan: we strive to inhabit everywhere, all at once. The globe becomes an artificial organism. The planet breathes as if alive once humanity has wired it for sound.

So Ramelli's wheel keeps draining the marsh, driven by the natural flow of the stream. Then the river is not current or eddy, but source of power. The soggy land becomes a dry place to build. In its drive to become more natural, and thus more conducive to building a human sense of the universe as a home, technology has changed what we see and call nature. Once we make something, we gain a practical kind of certainty which we then wish to project outward onto the world. So technology gives us explaining as well as extending power. As we develop machines that seem more self-sufficient, we imagine that the world itself is as precisely intricate and composed of parts as our own constructions. Once we conceive of nature as a machine, it is a short step to think of ourselves as machines as well. The next chapter traces the development of the analogy between the machine and the world, showing that the change in the sophistication of tools and machinery alters what is meant by the announcement that "nature is like a machine." This change leads to machines that challenge us to articulate what it is about humanity that is more than the record of our accomplishments. Our technology *does* propel us toward nature, while continually changing the image in the scope that sets the course.

The striking thing about this machine, a two-person well to force water out of a deep shaft, is its value as metaphor. On the surface of this earth, where the man and woman work to fill the jug, there are trees, grasses, and flowers blooming. But deep inside the ground is a hidden network of gears and rods, clockwork in motion. This is nature uncovered as technology, and it is through the success of devices like this one that the world begins to make sense as a mechanical apparition.

It has *always* been possible to look at nature as a machine. But the meaning of the machinery changes consistently through history. What we know how to make has always had a profound influence on what we claim to know about what is larger than our powers—this enveloping world which we want desperately to be able to call our home. Is this goal an illusion, some human lie which we trick ourselves into looking towards, all the while conveniently redesigning it to be compatible with whatever technical expectations we have? The last chapter demonstrated that many of the most influential attempts to place technology in the sphere of human affairs have described it as the complex that enables us to approach nature. A closer examination of these theories revealed that nature does not retain any simple meaning from theory to theory. As ideas and technology change, the features of nature change as well. Nature is unified only in that it remains a place in which we wish to be accepted, ever beyond the range of our knowledge.

Enzensberger writes a poem to depict the construction of nature through the viewing platform of technique. Describing the efforts of Eadweard Muybridge, one of the pioneers of moving pictures, he wonders just what has been revealed when a bird in flight has been captured on film:

The pigeon, tied to the jib of a merry-go-round—
does it fly, or is it flown? The trail
of its wings is invisible: yet it's followed by,
pneumatically steering through a chaos
of tubes and drums, a steel spike;
quivering, it scratches the soot-blackened paper.
What writes and draws there, measuring itself,
is a hallucination, known as *Nature*.[1]

Mirroring the secret of movement, the new camera creates a nature. There is no movie without the mechanism in action. Nature has been depicted with more realism, but is more than ever a mirage.

The more powerful technology becomes, the greater its influence on explanation. The subject of this chapter is not how technology extends us through material actions, but how it redirects our thought. This is the flip side of the technical realization of human intent—intentions are themselves renovated through the success of techniques. The entire array of desires is transformed as we are seduced into analogy by the things we have built and constructed. Nature was considered to be the leading context of technology in chapter 3, but it is equally the *result* of technology. Human nature, as well as external, encompassing nature are both swayed by changes in technology. Sometimes they are equally conceived to be mechanical, while at other times one or the other possesses some soul or spiritual force while the second is seen as mere clockwork: mind over matter, or individuals as tiny fragments living out some overarching plan. The development of machinery suggests new ways of comparing humanity to the universe, some which estrange us, others which invite us.

It has been said that the decision to look at the world as a machine foreshadows the downfall of Western civilization. But what if machines could become the kind of organic devices dreamt of by Lewis Mumford? Or remained the human-expanding tools idealized by Plato and Socrates? The word *machine* can mean many different (but all human) ways of interacting with the world. No swift definition separates it from the more directly encountered "tool." Those more elaborate and meticulous require more compartmentalized thinking to realize their purposes, but they still function only as human extensions. When we adopt a technology, we do so because it works, fulfilling specific intentions. The intention to explain the world is never realized so clearly, but it is natural to base such explanation upon those things that work more reliably and less ambiguously than free-floating ideas. The machine is accepted only when fully under control. To place nature under our rule, we think of it more like a machine. The same thing with ourselves, as our internal states can be solved as problems only if we, too, work along rational principles. The machine as exemplary method becomes a powerful guide to grasping systems which we are *unable* to construct. Yet the machine is not a static concept, and so our ideas about the world also move.

In this chapter we examine several epic stages of the machine, and how each affected its contemporaries' ideas of nature and human nature. From bow and lyre, craftsman's wheels, on to the clock, through the steam engine, and culminating at present with the digital computer, the changing mechanical appearance of nature will complement the quest for nature of the previous chapter, illustrating how technology lets us see only what it is able to see. Nature becomes only what may be dealt with by the pragmatic schemes of the techniques we employ to see it.

HANDS ON THE LATHE OF HEAVEN

When Heraclitus called the world bow and lyre long ago, it was more than an attempt to evoke the richness of nature with colorful language. This was the formulation of a particular rule in philosophic thinking: start an explanation with something the public can understand and accept, and build one's thoughts from there. Both these devices work upon the principle of tension and release: one holds and releases an arrow, the other sounds a musical note when plucked. Anyone who has tried out these tools will grasp at once the principle of their operation, if not the skill needed to use them well. When we then jump to say that the universe itself is connected upon principles of tension and release, we are already imagining it to work like a machine, because it is in machines that we have first seen what these two forms of energy, potential and kinetic, can accomplish together. But the universe remains infinitely more complex than the source of this analogy; daring to point out such a similarity sets our thoughts on the whole further toward abstraction.

With this in mind, no mechanical analogy ever completely replaces nature. The world is always greater than our collective interpretations of it. Yet the more impressed we are with the amount that machines can do, the harder it is to conceive of aspects of the world beyond their logic of operation. It is easy to forget the limit inherent in any attempt to explain. The success of mechanisms shows that the embodiment of ideas put together piece by piece can be used to realize desired ends. How convenient and comprehensible if the universe is put together the same way!

Dividing its material into component parts need not be the most salient feature of a machine. Consider the kind of *techne* most respected in ancient Athens, where so many of the explanatory dogmas of Western philosophy were born. Plato's ideal craftsman might be a potter, shaping her product upon a wheel, spinning around an axis, gently molding the finished, symmetrical form with the direct pressure of the hands. The potter's wheel, too, is a machine, one conceived as a flowing process whose discrete parts are only meaningful if they make the seamless stream of human creation possible. It is no surprise that this smooth, flowing example of mechanical device is able to suggest in tandem the Greek ideas of a unified, symmetrical universe and the possibility of a perfectible human being who may be shaped and formed upon the spinning, harmonious world.

Along with the potter's wheel comes the spindle that turns raw fiber into thread and the lathe that enables us to carve symmetrically into wood. These three familiar devices of the Athenian world display the extending significance of the wheel as human invention. More than a symmetrical object that rolls, it is the basis of tools that let us explore the discovery of symmetry. Once we can build upon order, we come to see the wonder of order as an organizing principle all around us.

These shaping, dynamic and humanly guided wheels are the roots of a form analogized from the hand of the craftsman unto the rule that is nature. The final image of Plato's *Republic* is the whirling universe revolving about the "spindle of Necessity,"[2] turned incessantly by the Fates, presiding over the blameless passage of souls from one life to the next. The vault of Heaven spins as a sphere, and each of us is folded somewhere into the woven cloth. There are no mistakes. In the *Timaeus*, Heaven is a lathe upon which the Creator has turned the world in the form of a globe, "the most perfect and the most like itself of all figures, for . . . the like is infinitely fairer than the unlike."[3]

The ingenious wheels connect human weavers, spinners, and turners to the possibility of perfect shapes that we are unable to make. We may create things, so it is possible to imagine a creator who could construct us and our world. Creation itself is a human activity before we analogize back to the precursor of the world. If we are to imagine that nature is an object created, it comes one step closer to being the kind of object people may possibly understand. Start, like Goethe's Faust, with the *Deed,* something we can reenact, unlike the elusive Word or Light. If the craftsman can produce symmetrical objects and reveal the beauty in specific pieces of reality, then we may fathom the creation of the swirling, perfect whorl of the whole, shaped and coaxed by some ultimately powerful, guiding hand. The power of the idea of an eternal circle is made much more emphatic when we know we have the technology to approach it with our bare hands.

As the cosmos has been smoothed, sanded, or spun toward perfection, so might a human life or society be similarly formed. The cosmologist inspired by the successes of craft need not say that humanity is distinct from nature, destined to follow a different plan. This model of universe in harmonious revolution is suggestive enough to be scaled down to the smaller human sphere. We are a lesser container in a Platonic series of nested boxes, all hierarchically placed in the context of the turning, dynamic whole.

The image of the universe and humanity as this kind of spinning machine possesses another kind of less obvious wisdom. Here are tools that directly extend human grasping and turning movements toward the production of round and symmetrical objects, more perfect as embodied than anything we could carve without the wheel. But they remain technologies *driven* by human intention, even as their importance shifts from means to construction, from direct action toward abstract thought. Most memorable in this ancient Greek model is not the picture of an independently spinning Earth or soul, but the idea of the *hand* perpetually turning the wheel. This is a vision of human extension spread out across the land—a notion of universe that knows it is no more than someone's idea. It could never detach the natural from the human, as it sees perfection only as something molded and shaped by some kind of guiding grip.

These tools which eke symmetry out of a turning motion have such analogical power because they seem to straddle several of the categories of technology introduced in chapter 2. Though the potter's wheel may start as a simple hand-driven tool, extending human circular motion, it quickly finds its application by allowing us to realize an abstract idea, perfect roundness, through the construction of vessels and pots. Did the idea of the circle intend the bowl, or did the fairly round pan inspire the reality of perfect circularity? Plato suggests that we reach from our imperfect materials toward the crystalline inner ideals of the mind, though he asks us to simply accept that the purity and ineffability of the ideal is somehow "better" than the tangible roundness of the thrown clay pots. Yet only the latter will hold water.

It does not really matter which came first. What is important is that the turning wheel graphically illustrates technique altering intent. Once we can make fairly symmetrical containers, we quickly think of new applications for such forms. Clearly the abstract beauty of the circle becomes more concrete if we can approximate it easily upon the wheel. In addition to increasing in practicality, its strength as an abstract principle grows. Only now it is that much nearer to us. The perfect vessel of nature may be thrown by someone with even more power to spin than humanity.

Conceptually significant tools not only suggest new uses for themselves, they offer working models for processes outside the extent of human prowess. To inspire such daring extrapolation, these devices must revolutionize a part of the human realm as well. Spinning tools bring prediction and regularity within the range of all who can turn a wheel and shape wood, clay, or thread upon its motion. They make order accessible. Knowledge and the course of the hand behind nature are meant to do the same.

So the most important technologies may be those that suggest new uses for themselves, and also imply new ways the world might work unto *it*self. We learn to see only through mastering the manipulation of basic natural forces. Coaxing symmetry out of rough material is a powerful testament to the connection between humanity and order. And if the universe too is ordered, we are closer to it as well.

Greek civilization was by no means ignorant of machines more complicated than these. Hero of Alexandria depicts wind-powered pipe organs and elaborate automatic fountains in his *Pneumatics*.[4] And the remarkable Antikythera device, found in 1900 in an ancient Aegean shipwreck, is supposed to be some kind of intricate astronomical calculating instrument, with gears and dials far more precise than those previously associated with Athenian civilization.[5] Even more perplexing than the existence of these intricate constructions is the fact that they did not seep into philosophical thinking the way the lathe, spindle, and wheel did. They may have been considered as curiosities because they were closer to self-contained mechanisms, unable to immediately analogize to a human in connection with the surrounding world. Technology that visibly extends us in a transformative way exerts greater conceptual power than that which acts automatically, imagined to be alive. Automata were known as toys for centuries before they were applied to pragmatic and explanatory tasks. The direct image of the tool in the service of craft proved hard to surpass, for a nature that allows us to fashion the forms we have discovered will not require our exclusion.

We are left out when we begin to see things which cannot be found and thus have to be built. An order imposed on the world, rather than shaped out of the common interaction of mind and material, simplifies the image of a universe finished by a trained hand into a model of grid-like punctuality. Nature is then set by the sweeping idea of time.

TO TEAR THE DAY TO SHREDS

The clock, like the oracle of Delphi, neither speaks nor conceals the truth, but *indicates*. It makes the regularity of nature into a dimension that we can never prove. There is no observable cycle as exact as our unwavering scale of hours and minutes, and nothing happens in a milli-second until we place our measure upon the surging stream of events. Here is a technology that does not *do* anything remarkable to extend human physical powers, nor does it heighten any recognizable percep-tual sense. Yet time, once it is imagined to be regular and predictable, opens up a whole new axis of perception beyond the constraint of sense. Trust in its reliability allows the construction of machines far more pow-erful than the direct extension of any human faculty. Upon its discovery, we are able to remake the world with assurance that planned change may run smoothly, and take exactly as long as it has always taken be-fore. Strange then that the unmistakable tick that we take to be time cannot be felt in the world before we have placed it there.

The clock, in the scheme of categories, is an embodied construction that runs on its own, needing only an original instigating push to set it in motion. If the timepiece is working properly, it is a fact of our envi-ronment, a device that constantly displays one precise kind of informa-tion, the correct and unquestionable occasion of the moment we choose to glance at it. But this is not the side that most alters the context of intentions. The clock is the embodiment of an idea, and whether driven by water, a pendulum, a crystal, or a chip, its significance lies in its im-mediate demonstration of the power of human intent to order the world upon faith in regularity. Nothing can easily challenge the homogeneous passage of time, because the perspective it places on all events is so sim-ple and so global. If time is uniform and ubiquitous, we can compare any two disparate instances to each other, using a criterion that lies *outside* experience. By measuring them against time we make the uncertain flow of our lives into something predictable.

What the clock indicates, however abstract, simplifies human life and nature by putting them on a schedule which need not be doubted. Time as measured marks all that may happen to us, though we still cannot locate it. Kant was wise enough to recognize time to be a human attribute, but he was too quick when he assumed that we could not possibly perceive the stream of events without complete faith in its recurrence. He forgot that human life was not so punctual until the installation of the clock as a tool to standardize the times of prayer in the medieval monastery.

As much as the abbeys were seats of devotion and Christian faith, sealed off from the supposedly barbaric practices of the surrounding world, they were also the primary centers for the preservation of human knowledge, and with this, the love of order which inspires us to care about the truths of science even if they do not directly enhance our lives. With religious and Scholastic discipline came a life of prescribed devotion, and as early as the seventh century, Pope Sabinianus decreed that the bells of each monastery should be struck seven times in each day, in order to mark the exact periods for prayer.[6] The formality of this style of life is enhanced with a technology which can remind the monks of their duty at the necessary moments. By the tenth century, refurbished ancient water clocks were used for this purpose, and in the years that followed, the mechanical clock and its recurring bell began to be heard far beyond the walls of the cloister as a pealing reminder of the hour. It became more of a comfort as the listening worker in the field or workshop came to accept it as familiar. (Is this like the digital watch that offers a personal beep every hour unless we figure out how to turn it off?)

Once the passing of the hours is taken for granted, it is possible to conceive of our lives as passages through the day of signposts and evenly placed gates. We are supposed to work for four hours and then take exactly one hour off for lunch. No more, no less. This life is considered efficient only because it is more predictable, closer to a discipline which we uphold as an ideal. No longer need we consider the ideal of order to be the form which craft can only approximate. Instead, it becomes a factor we may set our activities by, an abstraction that makes the flow of our thoughts and deeds set and concrete—*like clockwork*. A machine that counts unflinchingly, marking hours, then later minutes, and seconds without fail proves to be a device through which we discern the outlines of a self-contained, patterned universe. The mandate we yearn for is no longer patiently held back at eternity, but is present in the here and now, if we look for it with the right gauge in hand, or on the wrist.

The immanence of this order was powerful enough to quickly enter the discourse of analogy in the elucidation of both natural and human character. Under the spell of the clock, nature as machine is no longer a craftsman's tool, gently coaxed through its movement. Instead, the supreme Watchmaker need only set the construction in motion, as it is built according to rules that are already clear enough to ensure an equal, even-beating tide. In the fourteenth century, Nicole Oresme found in the clock a complete and self-contained model of the universe:

> The situation is much like that of someone making a clock and letting it run and continue its motion by itself. In this manner did God allow the heavens to be moved continually . . . according to the established order.[7]

The machine we call the image of nature is no longer something driven by human control but a device set up by human ingenuity to run its own course. We are able to imagine such a nature working independent of any human or superhuman guide. If the world in itself is as orderly as the kind of time we have applied to the pattern of our own lives, then we will be able to study and investigate it as we do the mechanisms of our own invention. The only difference: some machines we make, others we discover. But their operating principles are the same.

Evenness, standardization, predictability, uniformity, and analyzability into discrete parts that serve to enact the regular motion only when put together in the correct manner—these are the attributes of a clockwork nature, the penultimate machine in the age of the marking of hours. How much can we see within this rubric? This was one of the questions asked by René Descartes, and his answer reinforced the fundamental existing objection to the inclusion of humanity within nature: if all of nature is mechanical, how can we be contained within it and still claim to exercise free will or possess a soul which links us to God, who once set the Great Clock in motion for perpetuity?

Descartes's *method* is certainly the philosophical equivalent of the craft of the clockmaker—to divide all difficulties into as many parts as possible, so that the solution might be elucidated piece by piece, removing all shadows left by doubt. All that remains of certainty is the thinking subject, and once he accepts the singular reality of mind, he needs to account for the pale resonance of the rest of the world. This becomes matter, or body, and *this* is the great machine, which we may partake of while still remaining aloof.

But thought is anything but mechanical, so for the rest of the world to be machine, it must be carefully severed from all impact on our senses. So Descartes defines matter and body as *pure extension*.[8] He does not mean the extension of any conceivable human faculty, but somehow the opposite—extension itself, the filling of space by parts that encounter one another without reflection or consideration. Movement within the realm of body is simply the "transference of one part of matter from the vicinity of those bodies that are in immediate contact with it . . . into the vicinity of others."[9] Empty, extended patches of matter interlock and shift in relation to one another. The meaning of the machine composed of such parts must come from without, and this is the function of mind.

The clockwork automata enjoyed as toys by the Greeks now merit greater attention. If they reproduce certain simple aspects of living, self-moving things, it is because life forms may be conceived as possessing the same nature as those imitative devices we have long known how to construct. The key difference is that the Creator is a better machinist than we are: the contraption fashioned by the hands of God is far more advanced than any which human ingenuity could conceive. But animals (and presumably waterfalls, swamps and volcanoes as well) are still mechanical, by virtue of Descartes's observation that we would be unable to distinguish a mechanical monkey from a live one, providing all the beast's features could be modeled by appropriate invisible gears and rods.

So need everything be conceived as a machine? All except one thing: the human thinking soul. As long as our prime certainty is that we are thoughtful beings who may learn from the universe, we cannot ourselves be entirely analogized as machines. The originality of human nature is defined by Descartes in opposition to mechanistic, purely extended nature. Only we (or the Creator?) possess the knowledge necessary to set mechanisms in motion. Only reason is the "universal instrument which can serve for all contingencies,"[10] while pieces of the machine are always specialized for one particular task as cogs in the mechanism of extension. No animal or aspect of nature may then be as free and diverse as the human spirit.

In this attempt to protect humanity from the detrimental consequences of life in a time-bound world, Descartes makes us its prisoner. His proof that animals are no more nor less than machines is only that spectral mechanical monkey, which could so easily fool us into taking it to be the real thing. *We* are the arbiter of the strength of mechanism; the blueprint of nature as machine is still our interpretation. This mechanistic hypothesis requires sufficient faith in our own understanding to say *we know enough* about what surrounds us to call it mechanical and ourselves reasonable. For a machine is a meaningless thing until it is complete. If the fragments do not concurrently form a whole, they may only find value as spare parts for some other, as yet unbuilt device.

If we are the only thinking beings in this scheme of things, then we are confined to a uniquely human road. The order which offers the possibility for the expansion of embodied machines is imposed on the Earth, not found in the sky. The thinker extracted by Descartes from the world is destined to see nothing but machines outside of her own narrow human privilege. And when we pause to look within, the only tool that remains at our service is the methodology of breaking the complex into simples and then piecing them back together into a newly complicated whole. The dream of sheltering the soul is doomed, as the expansive mechanistic metaphor soon spreads to all rivulets of conception. We too will need to be admitted into the classification of machines if mechanism is to be complete in its explanatory foray. For this to be smoothly accomplished, an enhanced sense of the machine will need to be invoked as inspiration. The clock displays order, while the living machine must be an order that can move itself. Within such a new mechanical community, natural and human will again be one. What Descartes tried to save will be lost.

The earliest responses to Descartes tried to retain some kind of duality between humanity and nature, trying several different strategies in the usage of the machine in the comparison. Spinoza's machine of thought is a schematic, geometric elucidation of the human path out from nature and back again through the right kind of knowledge, as noted in chapter 3. The dynamic part of this vision lies in the way we travel through the conception, rather than in the workings of the universe itself. Leibniz solved the problem by setting up two separate but similar clocks to represent mind and matter distinctly, both independently established at the dawn of time out of an original harmony. The two complement each other by running in parallel, with each event marked independently, both events set in perpetual place by the mind of God. In this conception, even free will would be programmed in advance.

This is the paradox which all advocates of mechanism in the mind must confront at some point. To get beyond it, more comprehensive and flexible machines are needed for analogy. Before covering the way developing technology expands and strengthens the metaphor, it is necessary to consider the historical tendency to unify humanity and nature through a single mechanical image. This is the culmination of the clockwork model, first fully articulated in 1748 by a physician, Julien Offray de la Mettrie, in a work emphatically titled *L'homme machine*. La Mettrie attempts to rid mechanistic philosophy of the Cartesian specter of rationalism, using what he considers to be a purely empirical method. What is remarkable is not how he avoids the duality so precious to Descartes—that is easy once you deny uniqueness to the soul—but rather the emotionally charged language he uses to make machines themselves sound *daring and alive*.

First, the human body. This is clearly mechanical as long as it recoils from the world in a manner innate and involuntary:

> Is it not in a purely mechanical way that the body shrinks back with terror at the sight of an unforeseen precipice, that the eyelids are lowered at the menace of a blow? . . . that the heart, the arteries and the muscles contract in sleep as well as in waking hours, that the lungs serve as bellows continually in exercise?[11]

We may observe that our bodies work in predictable yet unmotivated ways, while recoil can still be chilling. This much Descartes recognized. The difficult area of inquiry concerns the soul. How can this trait that lifts humanity above material nature find its way into the rubric of mechanism? With the image of a machine that *runs itself*. A device that may set itself into motion can be compared to the indubitable thought that directs our attention from within the gears and wheels of our mechanical minds out into a world that is already a construction of spinning parts:

> The soul is clearly an *enlightened* machine. For finally, if man alone had received a share of natural law, would he be any less a machine for that? A few more wheels, a few more springs than in the most perfect animals, . . . any one of a number of unknown causes might always produce this delicate conscience so easily wounded, this remorse which is no more foreign to matter

than to thought, and in a word all the differences that are supposed to exist here. Could the *organism* then suffice for everything? Once more: yes. . . .[12]

All it takes is a few more parts. *Then* the machine may run itself. In his exemplification of machinery, La Mettrie rarely mentions words besides *wheel* or *spring* which might help us to envision just what kind of machine he has in mind. At one point the "cerebral fibers" are said to be struck by waves of sound the way a harpsichord key plucks a string,[13] but we must extrapolate from the fact of clockwork to imagine constructions of discrete and linked parts which are able to play themselves. The human being is the machine which winds its own springs, and we must believe a machine could possibly do that before we may call ourselves one of their kind.

And before this conclusion must come a faith that organisms are completely understandable under mechanical scrutiny. If there is a purpose, or teleology to life which *eludes* quantification, it can never be replaced by even the most convincing of machines. This has been the stance of vitalism, a challenge to mechanism that has continued unto the present day. There has always been someone asserting that meticulous principles can explain all the important facts of life, while critics abound to retort that analytic, compartmentalized thinking will inevitably elude the essence. The historian of Chinese science Joseph Needham penned one of his first books to uphold La Mettrie's enveloping mechanism against an early twentieth-century defense of vitalism. His essay of 1928 is entitled once more *Man a Machine,* and he emphasizes that the universal coverage of mechanism amounts to a defense of science as the route toward overall truth. Attempts to deny mechanism in nature or the mind are the biggest threats to a science that wants to examine both:

> *In science, man is a machine; or if he is not, then he is nothing at all.* In science such an aphorism as that of Samuel Butler, "A hen is only an egg's way of making another egg" is absolutely meaningless; in science, egg and hen revolve together within the circle of Immutable Necessity.[14]

Mechanism is the understated corollary of science, the attribute that must be accepted if we are to believe we can take nature apart piece by piece, and then put together an understanding even remotely like the original. We are forbidden to use teleology to try to distinguish life from nonlife. As it cannot be quantified or measured (there is only an end and a beginning—no percentage of finality) the goal is of no scientific significance. The ultimatum of an infinite guide can never be put into numbers or sliced into parts. So a cogwheel will never bring us to any distance too far to measure. What is quantifiable is also mechanical. We can do no further measurement upon the infinite, after we have given it a name.

Needham is astute enough to realize that science, with all its conviction, is still a point of view: "To the critical spirit, science is seen to be a dream as well; it is only veridical like the other dreams, not veritable."[15] Science complicates the world, but places the manifold of parts within a modicum of order. And people love order enough to imagine that it is always there, waiting to be revealed once we possess the tools to dissect it. The greatest tool is an idea, one that heeds us to imagine clocks wherever we turn. Does it open us up to further depths of experience, or does it structure experience in so singular a way that we become blind to all alternate forms? Our answer to this question hinges on how much we trust in the ubiquity of the machine. Of course if we accept it in one case, it is likely to bump and grind against enough other parts of nature to claim credence elsewhere. One push sets the works in motion. Yet can this shove come from within the system?

To summarize a hopeful view of technology's progress: the machine can only be salvaged as transcendent metaphor if it is *more* than a clock. We do not know what the clock indicates, and we must have a machine that keeps track of itself, one that moves and acts, rather than one so clearly limited to its own steady dimension of thrumming time. There is no reason to assume the evolution of technology will not continue to reveal mechanisms far more intricate than any previously known. Let Descartes have the last word in this paean to technique:

> We see that clocks, artificial fountains, mills, and other machines of this kind, although they have been built by us, do not for this reason lack the power to move by themselves in diverse ways: even respecting that machine which I suppose to have been made by the hands of God, it does not seem possible to me to imagine so many kinds of movement, nor to attribute to it so great an artifice, in such a way as to prevent one from thinking that *there can still be many more.*[16]

ENGINES, FUEL FOR THE MIND

A world seen to be clockwork is composed of tiny, interlocking parts that engage each other in a machine of relations, each like a gear whose smallest motion will turn all the other gears. If nature appears to be thus, humanity will inevitably come to be contained within the vision, as no impervious border between ourselves and the world can survive the probings of meticulous analysis. Our place is further defined the more we learn how our predictable actions alter the movements of the grand surrounding machine. Every motion has its exact effects; every action follows the precise dictum of an immutable beating time.

How far can this conception take us? It cannot explain why anything happens, nor how motion is instigated in the first place. If all mechanisms are like clockwork, then they are very much alike. One turn necessitates another, while nothing tangible is passed from one part to the next. It is hard for a clock to become conscious of its internal operation. This machine will never be aware of its own progress or decline.

Yet we want to believe humans possess such an ability to reflect, and that nature is able to move itself. Such challenges might threaten the metaphor of the machine, if it were not the case that machines began to evolve past the ultimatum of the clock. With the mastery of steam power in the eighteenth century, the *engine* becomes the technique with the most power to change the practical conditions of human life, and to suggest new ways to conceive the world and ourselves as parallel machines—still regular and dissectable, but now self-moving and voracious in the use of a primary resource, *energy*, which passes through us from piece to piece in a manner no clock could record.

Although the possibility of setting machines into motion with steam had been considered for several hundred years, it was not feasible until James Watt glimpsed a crucial insight in 1765 which could solve the problems of excess water and condensation: "The idea came to my mind that as steam was an elastic body it would rush into a vacuum, and if a connection were made between the cylinder and an exhausting vessel it would rush into it and might there be condensed without cooling the cylinder."[17] This sudden realization became the technical foundation of a whole generation of machinery and the industrial revolution that ensued.

The culmination of Watt's vision came sixteen years later with the patenting of the double-acting engine, with steam building up on either side of the piston, producing rotary motion from seesaw motion. As steam engines grew larger and more powerful, the alternating motion needed to be carefully regulated. The optimal movement of the pistons occurs only when the passage of steam through the opposite chambers is perfectly synchronized; so for the machine to function correctly, it needs to be aware that the gas flows precisely through the various parts at the right times. When the engine works, it is self-regulating, whether it knows it or not.

To explain the dramatic potential of his new device, Watt invented the measuring unit of horsepower (33,000 foot-pounds per minute, or 745.7 watts) to facilitate a comparison between his technology and the animal that it would soon come to replace—here the machine is measured by reference to a unit in nature. The first engines were 10 horsepower, and soon after, models up to 80 horsepower were developed for a large number of applications.[18] The reminder of "horse" became only a name, as the power produced by the bigger engines grew to be staggering in comparison. As we overcome nature, we begin to recast it in the new image of a mechanism that extracts, transforms, and uses up energy. Heidegger considered this to be the founding idea of the modern age, because it leads us to call all the environment a resource, as discussed in chapter 3. But in addition to refocusing our pragmatic gaze, it increases the power of our conceptual choice of mechanism to explain all.

The engine is, like the clock, not the direct extension of any move of the hand. When we set it in motion, it may independently perform tasks for us because it works upon the conversion of energy from potential form latent in coal and water, through steam, into a kinetic form that fulfills specific intentions. It differs from the clock in that it does something concrete, rather than just fixing the world according to a rhythmic, constant framework. Engines enable us to accomplish our ends at a distance, as all we need to do to keep them running is to guide the conversion of energy and pilot them towards the goal of our desires.

We set them in motion, but the machines take care of themselves, digesting the flow of fuel through their workings to keep the wheels moving. If the world works like these devices do, then we may see it not just as a series of interlocking gears, but as a collection of essential parts linked together by the flow of energy to realize a purpose. This is a far more powerful kind of machine in the realm of concepts, as it conveys a universe which is more than empty intricacy, beyond the playful automata of games. Engines use up material to get somewhere. Is it the same with humanity or nature?

If we are engines, then simply feed us the proper fuel, set us in the right direction, and we will do what we are told. If nature is an engine, it may be stoked by the shovel of God, but once set burning, it will run to perpetuity, as long as its fuel supply lasts. The clock is monolithic, a single abstraction subsuming duration under its chime, while each en-

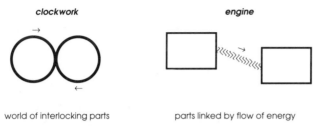

Figure 4. Clocks versus engines

gine can be self-contained, burning its own fuel, running in its own sweet time.

Potential energy ends up as a mix of motion and heat. How can the engine keep track of itself so that its smooth operation is insured? *Balance* is the norm, but only an equilibrium that is able to accomplish something with the smooth swallowing of what it is fed. The engine metaphor lasts well into our century. Norbert Wiener, pioneer in the field of cybernetics, explains succinctly the concept of an "organic system," developed in the nineteenth century to explain life upon principles tested in the eighteenth:

> The living organism is above all a heat engine, burning glucose or glycogen or starch, fat, and proteins into carbon dioxide, water, and urea. It is the metabolic balance which is the center of attention; and if the low working temperatures of animal muscle attract attention as opposed to the high working temperatures of a heat engine of similar efficiency, this fact is pushed into a corner and glibly explained by a contrast between the chemical energy of the living organism and the thermal energy of the heat engine.[19]

Balance, burning, work, efficiency—animals are not the only things that may be elucidated by these attributes of the engine. Volcanoes and earthquakes may be seen as storehouses of potential energy unleashing themselves as heat and movement, those very same sources which humanity may harness to fulfill its own perpetual need for power. And do not think that *we* are exempt from the descriptor of the engine simply because we possess minds or souls! Considering ourselves to be among the machines that consume, convert, and accomplish makes our lives seem even more prescriptive than if under the spell of the clock. If our food is conceived as fuel, consider how much clearer our lives may seem. This is a powerful marketing metaphor, which came to fruition as advertising grew to increased prominence as a means for refining our conception of ourselves. Consider this excerpt from an issue of the *Ladies Home Journal* in 1921:

> *FINE FUEL for YOUNG ENGINES!*
>
> Those young human engines—with their healthy, hundred horsepower appetites—what heaps of Aunt Jemima Pancakes they do consume! . . . And it's well that they do, for their ever-active little bodies need just such breakfast nourishment as they get in this famous food. It gives them an abundant store of energy to start the day.[20]

The image of stoking up our children with fuel at dawn is enough to suggest an image of the modern family as a unit built upon *efficiency*, Mumford's favorite buzzword from the Roaring Twenties. It was in this era that the machine was thought to be a tremendous boon to society at all levels, after the early indecencies of the industrial revolution had been ironed out. Technical progress meant improvement of life choices for all, and the future looked rosier than ever. We even believed we had healthier, easier, and better engineered foods to eat for breakfast, with which nothing could go wrong.

Of course, the penetrating image of machinery spread much deeper than this. Any thought about the actions and perceptions of humanity could not help but be influenced by the prevalence of analysis of experience into parts, each part a system that transformed some kind of energy. We had built a world that ran according to these principles, so it was natural once again to imagine those things beyond our power to create to follow the same rules. Michael Pupin, a reknowned educator at Columbia University in the twenties, urged everyone to confront their responsibility to see themselves as part of an unavoidable reality of mechanistic systems which "enframes" all the world:

> Remember that an incomparably larger number of machines are set in motion whenever a message of our sensations is transmitted from a part of our body to the brain. . . . When the organs of sense are busy transmitting our sensations, each one of their functional units is busy contributing its share to the performance of the organ. Each one of them is a machine and works at the expense of the energy supplied by the foodstuffs, just as *ordinary* machines work at the expense of the energy of wood, coal, gas and oil. We speak of feeding the fires under the boilers of our steam-engines just as we speak of feeding an organic body.

This mode of speech expresses the similarity between the two processes.[21]

What we build are "ordinary" machines, so the machinery of the body and mind are somehow extraordinary. But they are machines nonetheless, in a sense beyond mere metaphor. The perceived order of the world, including humanity, increases as we take it all to be mechanical. It is hard to resist this, and the difficulty is increased as machines extend their own reach, proving more useful in practice and analogy as a growing range of processes is mastered with analytical models.

What we lose in this rush to mechanize our conceptions is the very ability to accept what cannot be quantified into our vision. It is no accident that concurrent with the early twentieth-century rage for "machine-speak" was a movement in favor of vitalism in philosophy, led by Henri Bergson, which hoped for the articulation of a vision of the life force within nature which might also encompass human progress. The machine remains unable to tell us why we are going where we are going. Mechanism cannot direct or predict change. And yet we continue to change the world as our evolving technology alters the way we live in it as well as conceive in it. As long as this continues, *something* in our lives must evade the umbrella of the mechanical.

Joseph Needham already warned us: in science, humanity is a machine if we are to be anything at all. And we wish to be as suitable a scientific subject as the nature which surrounds us—how else can we fit into this world and make the recurring dream of home a reality? Bergson's influence in philosophy is generally considered to have faded precisely for this reason: science did not evolve in the direction he thought it should. No empirical evidence for *élan vital* could be found.

Science may have missed it, but there is a sense in which technology did find a way to mirror the living surge of the vitalist's dreams. The solution begins with the steam engine's ability to manage the flow of energy throughout its parts. For the engine to spin continuously, an internal, moving balance must be regulated. As the machine learns to keep track of itself, order is maintained in the pursuit of forward motion. A primitive sense of *feedback* is born. It is here that the methodology of cybernetics receives its initial inspiration, long before the electric and then electronic innovations that brought it to fruition in the form of the computer. Only a machine that keeps track of itself can gauge its own progress. This possibility makes thought conceivable in a totally mechanical world.

THE MACHINE STOPS

The second law of thermodynamics is a principle of the engine age. Given a closed system held together by the flow of energy, the rule is that order will always give way to disorder in the ceaseless path of time. Entropy is always on the increase. Organisms all eventually wind down. All engines are imperfect, and none runs forever. If the universe is the Ultimate Engine, then it too will expire someday.

Quite a pessimistic vision, yet nothing which could stop progress. An engine that can keep track of itself and respond to changing conditions is at least able to stay in motion awhile longer. The more the machine can assess its own state, the more successful it might be in the furtherance of movement. Devices that can do this are called feedback mechanisms, because the result of movement or change may be fed back into the instrument as fuel for a new process or decision (once more a mix between the organic eating metaphor and the mechanical notion of mutually influencing parts making up the whole). The alternating chambers of the steam engine are the simplest example, and they make possible more intricate instances like the engines of large ships whose sole purpose is to stabilize its position in the deep sea while other engines work on the problem of pushing the boat forward.

Do our muscles and minds work the same way? For computer pioneer Norbert Wiener, cybernetics could be the building block of all comprehensible processes, organic and inorganic, instigated by human or natural intentions. He considers himself the true heir to vitalism, having shown that the "modern automaton exists in the same sort of Bergsonian time as the living organism."[22] The only way to do this is to make better machines, and once more to generalize from the built to the contextual. What is added to the flow of energy through the engine is the ability to flow back around in a circle. Wiener sees in this model the root of a purely mechanical explanation for purposeful behavior in organism as well as device (fig. 5).

This development immediately extends the appeal of mechanism for theories of the mind. Without it, the vision of technology realizing intentions and then transforming them would be inconceivable. Cybernetics thus prefigures the thought process behind the present work. It is equally anticipatory of the notion of the "hermeneutic circle" developed by Hans-Georg Gadamer to articulate how human interpretive visions of history continue to cycle around each other, revolving about the truth.[23] Technology at last *encompasses* the circle, which was initially only es-

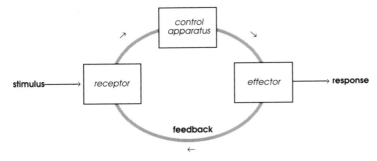

Figure 5. The cybernetic system (after von Bertalanffy)[24]

timable by the skilled hand of the potter smoothing clay upon the wheel. The cybernetic model encompasses roundness within its very construction, conceptually far deeper than symmetrical gears or wheels of any physical kind.

If grasping the wheel enabled us to imagine cycles in nature to be round in some way like the thing in our hands, the invention of self-regulating machines finally allows us to capture such cycles within the net of mechanistic thought. Only with feedback is a science like ecology possible, which is at last able to quantify systems in nature which would make little sense as clockwork. Or an empirical psychology, which can consider how individual humans react to precise "stimuli" from the environment, rather than the overall experience of living in the surrounding world.

At last, any organism can be divided into individual components that take up the various discrete functions defined as primary by the competing philosophies. A "receptor" takes in the sense impressions so elevated by David Hume, while a "control apparatus" is programmed to deal with these impressions according to definite, rationalist ideas like those guarded by Spinoza and Leibniz. An "effector" acts like a pragmatist in the actual world, and the one principle of feedback is enough to insure that these three independent units can smoothly keep up the flow. So the strong points of competing philosophical outlooks are synthesized in a system that gets results.

Once more, Wiener and his followers believe that they explain how technology and technical thought have come *closer to nature* by finding a way to include the seeking of goals within their actions. Being stuck in a feedback loop is the constant circulation towards an invisible goal. That end is homeostasis, a stability never glimpsed, but only realized through continuously striving from one immediate stage of the circle to the next. What clearer image could there be for the human predicament? Cybernetics shows us that all we need to do is to grasp the next and most immanent step, and because we are part of a cycling system, our motion has a meaning. We believe in fate, but will never see it.

Input, processing, output; groups of connected processes as self-contained systems—these notions have so worked themselves into our present-day thought that the word *cybernetics* has almost completely floated out of usage to mention their origination. It is too imposing and technical a term for a tendency in thought which has grown ubiquitous in our attempts to explain entities, individuals, communities, societies, ecosystems, planets, and the plan of nature itself. For Wiener, recognition of the system governing any situation is the first step toward elucidation of the vital force. *L'élan vital* is precisely what flows through the feedback circle that links a system's components. In the Steam Age it is the flow of energy which does it, but as the systems we isolate out of reality grow larger and more complex, the content of the flow needs itself to be more discrete and more complicated. Each component connects to the others via the interpretation of messages, received from the previous, passed on to the next. The content of the message is *information,* and the cycle becomes a path of communication. Energy can be called information, while entropy dissipates as lack of information. Order now means the finite edifice of data, and this will eventually run out, leaving the system with nothing. All rivers of facts will one day run dry.

Information quantifies the force of life. But notice how it dies as we say that. How could life actually be some series of numerical values thrown from one box on a diagrammed circle to another? As we take it as a fact to be analyzed, the living comes to resemble its image more and more, because the illustration teaches us what to look for, again making the model more than an analogy. It even challenges the vision of technologies as extensions. If devices tend to become "black boxes" that receive our input only to produce and spew back an output, it is hard to conceive of them as vehicles "driven" by human intention. Like the conceptual framework that they suggest and that makes them possible, these independent machines interpret our actions as signals to perform set tasks, with no need of a person to guide them. They watch themselves.

What they do extend is something far more subtle than a human intention to drive or to move. In line with the cybernetic circle, these embodied machines are extensions of the very idea they give rise to. It is impossible to talk about them without being drawn into the swirl. There is no need to speak of the computer yet: before the electronic age is the electric one, and this is where communication first happens with voltages passing through wires. The thermostat is the most oft-cited example of a simple feedback loop; as an on-off switch programmed to act at an exact temperature, it regulates the climate without needing to think too much about it. All it retains is the memory of a constant, and the ability to perform one specific action in response to the signal of the temperature reaching that point. What has been extended is the wish for a warmth more conducive to humanity than natural variation. The feasibility of a machine whose sole purpose is to decide when to turn the heat on or off is an outgrowth of the principles that made Watt's steam engine a leap beyond those that preceded him. It proves that technology can work even if abstracted away from the process it wants to control. The thermostat does little work itself; all it does is monitor the heating system. Self-regulation is a human idea, and we extend it through the things we build. Circularity is not only reinforced, but spread out respectively into each process over which we gain simple control.

The engine suggests feedback, and new machines extend feedback into new situations. The ensuing world view suggests a theory of technology which also inhabits the cycle. There seems to be little chance for escape. Even today's attempts to separate organic systems from mechanistic systems are often drawn into the rhetoric of the latter, as the very notion of system is a concept culled from the history of technology:

> Whereas a machine is constructed to produce a specific product or to carry out a specific task intended by its designer, an organism is primarily engaged in renewing itself; cells are breaking down and *building up structures,* tissues and organs are replacing their cells in continual cycles. . . . All these processes are *regulated* in such a way that the overall pattern of the organism is preserved, and this remarkable ability of *self-maintenance* persists under a variety of circumstances.[25]

Structure, building, regulation, maintenance—these are nothing but attributes of mechanism. Can we see nothing more than machines anymore in the world around us? Two cautionary sides of the cybernetic vision remind us that it is not an unequivocal fact, but still a humanly perceived approach to the world. It is necessary to consider these uncertainties before proceeding on to the computerized world that appears to fulfill the idea of constructed abstraction.

The first is the distinction between closed and open systems. A simple feedback loop acts as a closed system, as it does not need to consider any information outside the bond between its linked-together parts. But is it really independent from the rest of the world? Where does the stimulus that starts it come from, and what are the implications of its response? Any mechanism built by humanity starts with our intention, with the desire to create the machine in the first place. Mechanical systems do not *want* to keep track of themselves; the teleology of homeostasis is only an invisible end. The "first cause" of the machine is always the human wish for the solution of a mechanical problem. No complete understanding of a system can afford to ignore this. And yet the very circularity of the cybernetic loop encourages us to forget the reason that initiated the device. It is even harder to remember when we envisage self-regulating systems outside of our own constructions. Is the stimulus still our desire for the world to offer evidence of order?

Individual systems are bound to be linked together. Must they consequently appear as subsets of successively larger systems, like a series of nested boxes? A clockwork system would require such a hierarchy, while a system of energy flow need not. A line carrying a message may connect one system to another, independent of rank or class. Yet in terms of the Grand System, we believe that energy is always conserved, never lost, never found, only transformed in a finite, closed vision. Does cybernetics do the same for information—make it into a fixed quantity, and then cycle it throughout nature with innumerable metamorphoses? The universe as sealed river of energy continues to admit a certain amount of vagueness and mystery, as there may still be forms of energy unknown to us, which we may yet learn to transform and take. But information is strong because it is simple. It begins with the on and the off, the yes and the no, the one and the zero. Nature and humanity tallied as such will permit no mystery when we have attained *its* end.

So individual systems need to be open, but they seem to require closure at the level of ultimate wholes. And each system acts as if closed, with only a thin line connecting it to surrounding worlds. Cybernetic thinking works best around containment, and has difficulty with processes that appear to run wild (though the systematic appropriation of chaos may change this). Ludwig von Bertalanffy, who, with his general systems theory, fought hard for cybernetics as basis for a unified, comprehensive science, stressed that his would be an organizing theory that could uphold the individual mind, revealing an organism, not a "cog in the social machine."[26] Though the mind remained the greatest challenge to be charted as a system, it would be his final hurdle. The cycling image only proves itself when it embraces humanity as well as nature, with neither blinded by the result.

This is the liberating side. On the other hand, remember the Second Law of Thermodynamics. The whirr of circling ideas will wear down. Wiener, who believed the analogy went to the farthest reaches of reality, was brave enough to carry his idea to its conclusion. Entropy is still the final end of any closed system. Information will trickle out. Humanity too will die. But not because we will fail, only because we are part of the cycle. Every portion is finite, yet open. The total case is closed. He does not mean for us to take this as cause for despair, but only as the fact that dooms us to a swirling fate. Like Camus's Sisyphus, we continue inventing and watching our inventions change our thinking, all to keep us eternally afloat, a task which we know to be futile. As individuals die, each component system wears down. Only an empty nature remains.

Has cybernetics thus quantified humility? The computer is the device that at last embodies feedback principles, with a staggering range of embedded cycles and loops. Must it too be turned off sometime, signifying the end of the faith that order can be mirrored by a persistent cycle of on and off, zero and all? Pascal placed us between these extremes, but now they become the building blocks of an artificial nature whose strength is that it can make the most abstract of limits into a rigid frame that models all knowledge which we feed it. Unfortunately, it eats information one bit at a time. It is human intention that has divided the world this way, imagining circular mechanisms that fit the particles into something whole again.

INTRICACY LOOKS AT SILENCE

The computer changes the world into a machine too small for us to see, whose parts we are unable to touch, whose circuits we cannot trace even with the tip of a pencil. It is composed of microscopic constructions of logic, all linked together so they may collectively imitate any human thought process systematic enough to conform to its unassailable rules. Here it advertises itself as macroscopic, in its claim to follow all those actions of the mind which can be clearly explained as a series of logical moves. Whatever we comprehend as rule-governed behavior can be performed on and by the computer. It *excludes* the tacit world alluded to by Ludwig Wittgenstein: "What we cannot speak about we must pass over in silence."[27]

Here is the machine that opens up an ultimate challenge to humanity. It says: "I am the embodiment of your system of logic. I show the promise of ordered explanation. Imagine the world and yourselves to function like I do, and your notion of order can extend to deeper layers of complexity than ever before."

The machine does not have to speak to tell us this, for the message comes throughout our usage of this new tool as we solve problems and attempt to simulate external processes. If we can *describe* the world in a certain way, the computer can play with our descriptions, far faster than we could ever manipulate them. So does it beat us at our own game? This is the wrong question. These instruments provide the most tangible extension of the most ineffable of human activities, the workings of the mind. But to build worlds out of ideas, we need to turn them into solid components, not fluxing thoughts. Humanity appears to be extended, though any intrinsic silence impervious to the shackle of rule stays untouched, and becomes increasingly difficult to see as we forget what it means to know something beyond the bounds of explanation.

The computer is the hardest device to classify in the scheme of technologies, because it can be seen from so many divergent sides. Physically built out of the principles that have conceptually formed the systematic side of our thinking, the digital machine is at last a device that retains pieces of memory, subject to a theoretical speed of recall far beyond that of books, films, or photographs. As it gains prowess in the repetition of those human thought processes which can be quantified, it seems to replace our judgment in carefully controlled instances. So in the end it lives as a separate reflection, working for us without any continuous guidance, a fact of our environment as much as roads and buildings. The computer appears as a tool both abstract and material, humanly driven and independent. It challenges any attempt to pigeonhole it in any one place.

Yet despite its glaring independence, and the intangible power in the circuits beneath its "black box" shell, the computer remains a crystalline example of a technology which only achieves meaning as human extension. Digital technology furthers human faith in certainty and preference for order over chaos. After dividing all processes into systems that may be broken down to the unambiguous binary numerical base, we reach the ultimate coup of analysis: all complexes mirrored as simples. Any system seen as such will be reconstituted as closed. Any openness retained will be of a precisely programmed kind; it will accept inputs, and it will provide outputs, only in the proper formats. The program that replicates a real-world system, either one built by us or seen by us, will not possess an active relation to its context outside set boundaries. It cannot act without expectation. Though external changes may be accounted for, it cannot doubt the conceptual framework upon which it is written. A human thinker always can. A natural process is always more than the system we impose on it. The world *includes* what we must pass over in silence.

This view is contrary to what some researchers in artificial intelligence say, claiming there is no fundamental barrier preventing computers from one day thinking in the same way humans do. I believe this to be a faulty expectation, one that mistakenly imagines computers to be comprehensible apart from human intention. Computers do present a formidable challenge to human thought, as they redefine what technology can mean. The integrated circuit has no moving parts, can fit on the tip of a finger, and is an immensely powerful machine. It no longer just suggests physical processes, but mental ones as well. Computers transform mechanism into something so subtle that it no longer depends on tactile analogies like wheels, spindles, chambers or wires for its models. They dare us to submit our own minds to a kind of analogy so extensive we have difficulty realizing its incompleteness. Does our mind make more sense when imagined as a series of chips, ineffably processing the information of sense into the information of act? Does the mind of nature follow similar rules in a kindred spirit? What is certain is that, after engaging with computers in any fashion, one's view of the world will begin to change, with no possible way back.

A digital computer is a machine that admits a finite series of states, as opposed to an analog computer, which offers a potentially infinite series, limited only by our ability to measure it. A slide rule is an example of the latter, or any device that manipulates continuous flows instead of discrete units of information. Calculations are possible with networks of flowing electricity or even water, but the inaccuracy of measuring devices prohibits these instruments from getting more precise than human ability to regulate physical forces. The digital arena is far more abstract, because it may be programmed in a virtual realm, with little correlation to actual material forces.

Aside from the abacus, the first digital adding machine is said to have been constructed by Pascal in 1642. It is not known how this invention affected his cartography of human infinities, though it is likely that the intention behind the device was to reduce human drudgery so that we would have time for grander, more spiritual matters worthy of our precious but limited time. A century and a half later, mechanical precision had become more important to European culture at large, which required more refined and exact calculations for the careful production of an increasingly intricate industrial technology. Logarithmic tables were prepared by large teams of skilled human calculators, a costly and cumbersome enterprise. The public dreamed of machines that could help, and inventors assembled their gears to attempt them.

The maverick British mathematician Charles Babbage constructed a small model of a "Difference Engine" in 1822—not an engine in the technical sense, as it did not convert energy, but only mathematical difference. The machine was more pointedly a form of clockwork, composed of numerous tiny gears and rods turning dials upon which integers could provide a readout accurate to several decimal places. Unfortunately, the intricacy of mechanism necessary to perform such a formulaic task was beyond the precision of the machinery of the day. The idea, though a technical one, exceeded the reality of nineteenth-century technology. Babbage's concept of the computer foreshadowed developments that would come to fruition only in the present century, when materials more precise than clockwork could be adapted to instantiate the rigor of logic.

The radical nature of Babbage's idea of the computer emerges with the conception of his second device, the "Analytic Engine." This more comprehensive engine was to take its input in the form of punched cards of the type that had been used successfully in the Jacquard loom, another early digital device, which could automatically weave textile patterns the way Babbage wanted to weave mathematical ideas. Although he had still not finished any of his previous mechanical innovations, Babbage realized that the potential within such a device would be tremendous: any problem that could be broken down into discrete functions could be worked out on a machine, providing that we could manufacture the clockwork fine enough for the purpose.

The corollary to this realization was provided by Babbage's partner Lady Ada Lovelace, daughter of Lord Byron, who was far more adept than Babbage at explaining the machinery they collaborated upon: "The Analytic Engine has no pretensions whatever to originate anything. It can do whatever we know how to order it to perform."[28] This oft-quoted credo of digital culture recognizes the computer to be an agent of humanity, theoretically adept at realizing the intentions of the programmer, the person trained to weave real problems into systematic representations. The machine only acts for us, and lives out our logic. As our ability to systematize increases, so does the imagined prowess of the machine. And when the computer becomes more than a dream, it reinforces the precision woven cloth of our minds, which has made it possible. What the machine may emulate enjoys more of our trust.

Babbage and Lovelace failed to finance the construction of the Analytic Engine, but they grasped the promise of digital problem solving long before it was possible to render it tangible through electric current and symbolic logic. More practical experience with the digital approach might also have taught them its drawbacks, or rather, its tendency to transform problems into material it can deal with. The digital answer is not necessarily certain, it is only simple. A cybernetic feedback loop like a thermostat is able to control the continuously changing temperature with a basic, digital on/off switch. But it does not act continuously. It responds at a precise temperature, turning the system on or off. Heating is a continuous, analog problem, which only becomes digital in the simplest sense of control. A device that monitored the temperature at all times, compensating all the while to maintain a precise temperature, would be much more complicated. The digital approach *samples* reality; it is at best continual, never continuous.

This realization betrays an inevitable imperfection in the digitization of the world, natural or human: continuity is destroyed. With its on/off, either/or simplicity, the computer is founded on an inherently finite kind of mathematics of technological, mechanical origin. Previous mathematics dealt with a realm of ineffable formality, which illustrations or constructions could only emulate, while algebraic and geometric theorems demonstrated the truth of an ideal world, which our physicality might at best approximate. The certainty we feel after having grasped the Pythagorean Theorem is more definite than any fleeting experience, or so generations of philosophers have thought. It is the appearance of this kind of certainty within us that led them to hope that equal certainty might be discovered in other realms, such as the moral and the spiritual.

But the world prefigured by the computer is more modest than that. It exalts not the perfect triangle but the switch, not an ideal shape but a tool that just says yes or no. There is no perfection, only successive approximations through rapid repetition of the quantifiable. We never grasp a computed truth in an instant, but instead through the whirring operation of the machine, manipulating bit upon bit at the order of nanoseconds, impossible for us to see, impervious to the touch. Our original vision of "natural laws" was more than this, wishing for absolute, regulating truths by which the universe abides. The empty programmability of the computer challenges us to put these laws into practice by letting us build a world which follows the rules. Could such a machine ever be stronger than the pieces of logic which guide it? The entropy of the Engine Age still prevails, and the tool will wear out; bytes will be skipped and be written by mistake. The digital machine will still stop. The primary power of logical explanation is its ability to change the world as it teaches us a new way to inhabit it. We chart its boundaries decision by decision, after the unequivocal true and false of the logic table or tree, devoid of content. Yet logic machines do assess the world; they free us from menial mathematics by showing us how to see the world completely in quantifiable terms.

So it is not surprising that the next major innovation in computing to follow the impossible mechanisms of Babbage was not an actual device, but a theoretical one. In 1937, a twenty-five year old mathematician named Alan Turing, not coincidentally an acquaintance of Wittgenstein at Cambridge, published a short paper in which he introduced the possibility of a contraption we now call the "Turing Machine," whose credibility opens the route towards the modern digital computer. He asks us to imagine a box through which passes an infinitely long tape, divided into cells of equal size. Inside the box is a machine that can perform three classes of operations on any single location on the tape: 1) it may move the tape one cell to the left or to the right; 2) it may write onto the cell on the tape one of a finite set of symbols, such as zero or one;[29] and 3) it may change the symbol in the current cell.

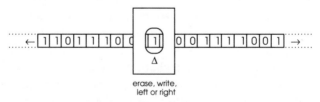

erase, write,
left or right

Figure 6. The Turing Machine

The Turing Machine performs no physical work as an engine does; nor does it mark any regular rhythm of the universe like a clock. It simply moves a marker back and forth along a tape, writing and erasing simple symbols sequentially. It *processes information* in its most fundamental form.

Turing's paper demonstrates that his machine is the most generic mechanical model possible. He proved mathematically that this imaginary device could perform any task that a special-purpose automaton could, because any mechanism must by necessity be detachable into discrete decision processes. Any delimited set of rules can be emulated on such a machine, and the logical consistency of this argument paved the way for engineers to create physical analogies to the initially imaginary process. Babbage acted on a hunch with his punched cards, but the computer scientists of World War II were bolstered by a logical proof that assured them that their mission stood on solid, if abstract grounds. Here is a thought model tailor-made for technology, as it sets up an acceptable challenge: create a machine that can mark ones and zeros on a linear path with tremendous speed, and I can show you how to develop rules to mirror any mechanical process. Turing justifies the search for an ultimate general machine, whose barren simplicity will make it adaptable to all step-by-step problems. Never before had a theory promised so much radical change in the meaning of the word *machine.*

The first practical application of Turing's imaginary reader and writer of tape was a project to break the secret code of the German High Command during the war, a cipher mechanically produced by the "Enigma" machine. Turing built a device to mechanically test out the thousands of possible algorithms by which a message might have been scrambled, using a trial and error method which would be too slow for any human team to attempt. Turing's contribution was instrumental to the British victory. A machine that began as a thought experiment helped to win a war. Any completely quantifiable repetitive task might conceivably be performed faster and with less error by a generalized automaton than by a human being or group thereof. Human ways of breaking codes or solving puzzles might be impossible to emulate by the logic of division, but a mechanical method was employed where our own unaided thought processes would have failed.

The next major advance, which made the physical construction of more generic computers possible, came from a new model of how the human mind was put together. We perform many tasks, not just one at a time, yet we have only one brain. The brain was understood as an assembly of components, with specific neurons or regions performing memory, sensory, control and motor functions. We had already adapted the language of machinery to explain our own organic minds, so why not build the computer upon the same principles? Here is a clear example of the cycle of extending technology, where the metaphor of the machine is thrown back upon our own selves, inspiring us to want to build an artificial brain that works in the way we imagine that we do. One machine suggests another: the link is the influence of technical success upon the search for explanations.

John von Neumann is usually credited with the introduction of this image into the world of computing, though there is some dispute on the matter.[30] He envisioned an enhanced Turing Machine divided into two basic parts, the central processing unit (CPU) and the memory. The CPU is like Turing's original machine, dealing with whatever problem is at hand by operating upon the data unit by unit according to an established program. What von Neumann adds is the idea that both the data and the program can be stored within the computer itself, in the section defined as the memory. It doesn't have to be a paper or magnetic tape, but could be tubes, transistor logic gates, and now circuit chips and electromagnetic disks of various forms, floppy and hard. The steam engine worked by recognizing coal as potential energy, and then providing a way to store it. The digital computer becomes a reality because it identifies binary data as information, and then includes within its closed system a means to store information as material and instruction. It is not used up like coal or plutonium, but evaluated like a mathematical formula.

With this model of the human brain which works because it is a linked assembly of functional parts, the seeds of an artificial thinking machine have been planted. If the Turing Machine proves the systematic nature of all thought, then the von Neumann model hints at how we might build systems that can think. Yet neither is the case. Turing simply showed that any rule-based process might be performed on a machine, and von Neumann suggested a way that such a machine could actually be built. Neither claimed that *all* of human intelligence could be modeled in this manner.

But for a machine to think, it does not need to replace all human thought. It need only perform even one action, or give even one indication, that it has done something worthy of the word *thought*. This is not without its own problematics. Is it safe to call a thought an operation, or does this already restrict a mystery to the kind of things computers have been designed to do? And how would we know that a machine is thinking; do we even understand enough about thought to recognize it for certain when it appears before us? As the most general and analogical of machines, it is clear that the computer will change the very notion of what thought is or can be. The operation of what we create still remains far more crystalline than the things we do not consciously design. If the computer is based upon a particular model of human thought, it will likely be a more accurate instance of that model than of the human brain, which we use and analyze, but do not build.

These difficulties aside, Turing proposed in 1950 another theoretical game by which we can investigate the possibility of intelligence.[31] Imagine the following scenario, which is today known as the "Turing Test": allow both a machine (A) and a person (B) to communicate to (C), a human interrogator, by means of the same mechanical medium, such as an interactive display terminal or personal computer.

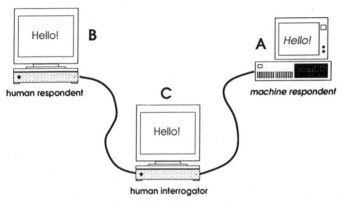

Figure 7. The Turing Test

Interrogator C might, for example, type in questions and receive answers displayed on the screen from either A or B. If C cannot tell which responses come from the machine A, and which from the person B, then they are formally indistinguishable from the perspective of C. If both responses are perceived as being intelligent, then both A and B may be called "intelligences." A wins the distinction, then, of *artificial intelligence*. Does this mean C has been fooled? Has he confused human and mechanical interlocutors? Yes, but only in the context of the game. If intelligence is what has been conveyed over the linkage of the terminals, and if B is not fundamentally more original, creative, quirky, unpredictable, or otherwise identifiably "human" than A, then they have been shown to be equally intelligent.

Has a machine ever won this game? The more poignant question is:

would a machine *care* whether it had won or not? The Turing Test is not really a contest between person and machine, but an entirely human challenge. It forces us to consider just how much we know about ourselves and what it means for a person to think. If we are tricked into believing that a person is on the other end of the line when it is in fact a machine, it is we who have lost. We have learned that we do not know as much about people as we thought. A machine has masqueraded as one of us, and we have been taken in.

The mistake is not that the machine should be recognized as an "other," impervious to our attempt to interact. Rather, it is a part of us, and our error is in mistaking a part for the whole. The computer is that part of us into which we have extended our symbolized and constructed logic, and no more. When we ask if it possesses a human attribute—"Is the machine intelligent? Can computers think?"—we are introducing a comparison. A human standard is invoked. We are asking, "Is it human? Is it *like us*?" It is not possible to isolate intelligence from other human qualities, detaching it from emotion, feeling, physical sensation of the body, or other defining yet nebulous hallmarks of humanity. Intelligence is always weakened when these sides of it are taken away.

The Turing Test is certainly not an empty game. It just teaches nothing of the notion of an intelligence apart from humanity. It is more of a check on how far detached quantification can go, compelling us to measure ourselves against the scale of rigorous imitation of rules we have been able to construct. It implores us to go farther, to learn more about ourselves. The game examines how we interact with the products of our labors, graphically revealing how an extension may look back on us and force us to consider how much we are aware of our own intentions and potential. No two machines could evaluate each other's intelligence. Only a human could ask the question or be persuaded to pay attention to an answer.

Turing himself, having abstracted the idea of the "machine" to mean any assembly of simple rule-following logical instructions, was able to curtly state the consequences. He was *not* in agreement with Wiener or La Mettrie: "If each man had a definite set of rules of conduct by which he regulated his life he would be no better than a machine. But there are no such rules, so men *cannot* be machines."[32] Yet he was wise enough to realize that the problem-solving flexibility of digital information processing systems would change our understanding of the concepts of thought and intelligence for ever:

The original question "Can machines think?" I believe to be too meaningless to deserve discussion. Nevertheless, I believe that at the end of the century the use of words and general educated opinion will have altered so much that one will be able to speak of machines thinking without expecting to be contradicted.[33]

If his prophecy is correct today, it means that we are now more apt to identify "thinking" with those processes which can be quantified and modeled on a computer with results that appear human, even if the processes by which they operate are totally different from the more fluid human ones. Computers can do an immense number of simple, unambiguous operations sequentially and fast. But programming them to synthesize and to consider many views of a problem at once is another matter. We are not sure how to quantify more complex models of thought, especially those flashes felt as insight or discovery.

When logic is extended from the caverns of the mind into the circuits of a computer, our conviction in it hardens. Logic began as a device to explain, and it has become a selective mirror which may reflect those parts of the outside world which fit the criteria it has been programmed to expect. Then we who have been influenced by the successful digital simulation of the world come to see the original in a similar manner. We become adept at ignoring what cannot be explained. Through a device that affirms the shell of rules which makes some behavior and thought predictable, we extend ourselves back upon ourselves. When the computer works, the power of regulation is demonstrated. We evaluate the prowess of a computer by comparison to ourselves. Intelligence is an anthropomorphic concept. It is never the machine that *wants* to engage in our games. We intend the encounter. We pass judgment upon it. Imagining that anything more occurs is so far a delusion, if at times a chilling one.

The most well-known example of a computer program that fooled many of its more trusting users into believing that it understood the sentences fed into it is Joseph Weizenbaum's *Eliza,* named after the character in Shaw's *Pygmalion* who learns to speak English better and better as the play progresses. *Eliza,* however, is a program originally intended to simulate conversation *without* understanding, like what goes on at cocktail parties when one pretends to grasp the essence of what the neighboring theoretical physicist or bond trader is talking about. You nod your head at the right places, say a few words that vaguely connect to what your interlocutor is saying, and if you act attentive, she thinks you understand her. *Eliza* does the same thing, by picking up on certain key phrases, or by spitting out abstract but leading remarks. A program of this form could be instructed to imitate many specific styles of conversation, and the most controversial was one that imitated a psychotherapist of the Rogerian persuasion, who specializes in "encouraging a patient to keep talking."[34] Here is an excerpt from Weizenbaum's example of a session between a young woman and the unknowing, but responsive machine:

> It's true. I am unhappy.
> *Do you think coming here will help you not to be unhappy?*
> I need some help, that much seems certain.
> *What would it mean to you if you got some help?*
> Perhaps I could learn to get along with my mother.
> *Tell me more about your family.*
> My mother takes care of me.
> *Who else in your family takes care of you?*
> My father.
> *Your father.*
> You are like my father in some ways.
> *What resemblance do you see?*
> You are not very aggressive but I think you don't want me to notice that.
> *What makes you think I am not very aggressive?*
> You don't argue with me.
> *Why do you think I don't argue with you?*[35]

The computer is programmed to pick up on certain key words or phrases, and respond by either repeating the word in an inquisitive sentence, or leading on to related areas, as from "mother" to "family." Otherwise, it simply spouts back what was said to it to encourage the "patient" to keep talking. Beyond this level of automatic response, it possesses no level of understanding of what it is doing. The reader may be surprised, and appreciate the ironic way the program is neither aggressive nor argumentative, but confrontational in an amazingly simple way. No computer would recognize the implications of the apparent ease in which a conversation can be had with a machine that knows only a few manipulative rules.

What exactly are the implications? What shocked Weizenbaum the most was the way the psychiatric profession reacted to what he had initially written as an amusing game. Some of them took it quite seriously: "Because of the time-sharing capabilities of modern and future computers, several hundred patients an hour could be handled by a computer designed for this purpose."[36] If even a few psychiatrists were convinced by this intentional parody of their work, what does it say about the profession? The first conclusion seems to be that psychotherapy can appear convincing if reduced to a few absurdly simple rules. Yet what is more significant is what *Eliza* reveals about therapeutic communication through the experience of those "patients" who have chosen to submit to it and then find they feel better afterward. Any success shows that the subject does not expect very much from the therapist, and benefits greatly from a chance to pour out his troubles to a listener who does not know the meaning of the words "boring" or "your time is up." Such therapy is a form of self-examination. *Eliza* reflects our input back to us with a twist.

Weizenbaum's take on Turing encourages us to take note of the ways conversation offers real engagement between unpredictable minds. It challenges us not to take such contact for granted. In this way the computer solicits us to question how human interaction is affected by technology. If *Eliza* appears intelligent to even one person, the Turing Test has been passed. But this tells us less about how much computers can do than about how little people expect from each other. This is how the realization of the machine should redirect our thought: it should lead us to return to a consideration of technology that opens up communication between persons and the world with a facile simulation. Another alternative, however, is a more likely outcome: we try to reconceive of ourselves so that we lie within the discourse of computing. Those psychiatrists who saw real promise in the clinical application of *Eliza* had to see themselves in an entirely new way: "A human therapist can be viewed as an information processor and decision maker with a set of decision rules, . . . guided in these decisions by rough empiric rules telling what is appropriate to say and not to say in certain contexts."[37] We venture onto this kind of course of self-examination after being caught by the computer's apparent achievement. We yearn for absolute order enough to want to reflect it back from the tentative machine into a framework explaining our own minds.

Weizenbaum himself was frightened enough by the devout seriousness with which his intended game was received to begin to question the discipline of artificial intelligence as a whole. He went on to advocate the value of moral training as a prerequisite to entry into the digital world. This is certainly something we will not find in the model of the machine, so we had best bring it with us. Initially called a traitor to the cause by his MIT colleagues, the significance of his doubt and cautionary response is now generally accepted.[38] The most profound message is that we really know very little about our own intelligence, and may easily be fooled by machines that emulate only the barest outline of our mind's ways. And as we apprehend the outline, we tend to reduce ourselves to those simplified structures whose correctness has been vouched for by the computer.

This is not to say that the only way computers have been able to masquerade as intelligent is to make a mockery of communication. If intelligence is confined to chiefly algorithmic problem solving, computers can be instructed to advance. The issue is especially provocative if the problem to be solved lies *almost* within the reach of rules, with an edge of indescribable human insight preventing the machine from obvious mastery. This is why teaching a computer to play the game of chess has so fascinated programmers. We know a lot of rules the machine ought to learn, but, like all artistic ventures, the secrets of play at the game's highest level are unknown. The flexibility and daring of the greatest chess players cannot (yet) be reduced to the ever-necessary set of axioms, so chess programs continue to lose to the world's best players. But they are getting better and better! If they do surpass the world champions, will there be no reason to continue to play the game? Will we be forced to concede to the computer at last?

The story of successive contests between humans and machines over the game of chess reveals why the computer remains, like all technologies, an extension of ourselves towards an uncertain nature. The content of a chess game may be contained on a machine, but the *appreciation* of the encounter never will be. The excitement of chess is an all-too-human feeling. Only we care if the machine will one day win.

Turing's automatic cryptographer cracked the Enigma code by trying out successive possibilities far faster than any human could dream of doing. Claude Shannon realized as early as 1950 that a chess-playing program could not work the same way. A computer that tried to calculate every possible variation a million times a second would still require over 10^{95} years to decide on its first move.[39] No human would have that much patience to play it. What is needed is a machine that can learn from its mistakes—one that could study and store existing chess games and investigate the probability of a chain of interrelated moves and patterns of play. Researchers worked on these programs throughout the fifties and sixties, and gradually improving performance was considered a barometer of artificial intelligence research.

The philosopher Hubert Dreyfus was bold enough to proclaim that computers would never be able to play a significant game of chess, as opposed to checkers, because the former involves processes too complex to be fully analyzed. A match between Dreyfus and the leading chess program of 1965, *MacHack,* was organized and, to the surprise of pro and con in the audience alike, Dreyfus lost. Another strike against philosophy? Not necessarily. Dreyfus admitted to be a rank amateur in the game, not a worthy opponent to test the depths of one of the most subtle of human intellectual games. *Humanity* and *subtlety* are the key words here. One has only to glance at Herbert Simon's account to realize that the match mattered more to the people who watched than to the machine that won:

> Dreyfus was being beaten fairly badly and then he found a move which could've captured the opponent's queen. And the only way the opponent could get out of this was to keep Dreyfus in check with his own queen until he could fork the queen and king and exchange them. And the program proceeded to do exactly that. And as soon as it had done that, Dreyfus' game fell to pieces, and then it checkmated him right in the middle of the board. *So it wasn't mechanical at all*; it was a typical game between humans with these great moments of drama and disaster that go on in such games. It was *wonderful*.[40]

What Simon enjoyed in his experience of the game is something the machine did not. Machines may win a few, lose a few. But it is not their part to appreciate even their own successes. This is because humanity does not know enough to program *care*. The game is always more than the rules of the game, as the lived value of play cannot be quantified. Of course we may try to exclude this from the domain of intelligence, but then we are left with fewer criteria to gauge the presence of thought as it happens.

The computer coaxes us into extracting thought from body the way the steam engine encouraged us to extract energy out of the Earth. As nature then appeared to be only a source for energy, the human being comes into the light as merely the source of thought. There is a chilling emptiness left by both kinds of detachment. Each development is a comment on the way we use the world.

By 1989, a computer program known as *Deep Thought,* a product of

Carnegie-Mellon University, achieved an international rating of 2500, as the first machine able to defeat a grand master in a tournament.[41] Later, the program was pit against world champion Gary Kasparov, whose rating is a bit above 2700. Kasparov effectively demolished the machine's line of play after only ten moves, making it look silly to the apprehensive audience. Kasparov himself described the difference between playing a machine and a human in this most telling way:

> When playing versus a human being there is energy going between us. Today I was puzzled because I felt no opponent, no energy—kind of like a black hole, into which my energy could disappear. But I discovered a new source of energy, *from the audience to me,* and I thank you very much for this enormous energy supply.[42]

The energy Kasparov wants to feel is not something that is extracted or transformed, but a wave of empathy which passes between beings who share some fundamental nature. He seems not at all impressed that the computer shares with him the craft of being able to follow the game of chess—probably because he noticed early on that *Deep Thought* does not share the deep involvement with the tension of each move that unites the best players. Chess is a game which is great because its simple rules test the limits of one human mind to predict the actions of another. There is no luck or repetition, only players' evaluation of each other's plan and commitment to a determined course. Each player respects the other because she knows what has to be gone through to learn the game. A certain range of experience is shared. But Kasparov knows that the machine plays chess in a manner wholly different from any human. It has learned the game in a way inappropriate for an individual mind. So he may be curious about the results, but is not touched by them.

Yet the match unites him with the audience. All who watch are human, all confronted by an alien who has been taught the rules of one of our games. The failure of the computer is a reminder that we do not fully understand ourselves, so we cannot beat ourselves at a task that exceeds the comprehensivity of rules. It is simultaneously an affirmation of and resignation to the limits of the human mind. As we consider the computer to be something separate enough from us to compete with us, we are all drawn together by virtue of our unquestionable, if elusive, humanity. We realize the computer mirrors only one part of us, and extends that part to whatever conclusions logic may offer. Kasparov has not followed in the footsteps of those psychiatrists who hoped for computer-assisted therapy. He has not let the computer make us seem more like it, at home in a universally flexible, if wholly determined world. No, the grand master knows that the machine misses an essential facet of humanity.

Although the computer does not imitate humanity, it extends a part of humanity which has previously eluded extension. This strangeness can alter our whole way of considering machines. The movement in favor of artificial intelligence is part of a larger tendency to encourage a certain human humility in the face of machinery of our own design. The following statement by Edward Fredkin, an extremist even in AI circles, presents the implications of this view:

> Humans are okay. I'm glad to be one. I like them in general, but they're only human. It's nothing to complain about. Humans aren't the best ditch diggers in the world, machines are. And humans can't lift as much as a crane [or] carry as much as a truck. It doesn't make me feel bad. There were people whose thing in life was completely physical—John Henry and the steam hammer. *Now we're up against the intellectual steam hammer.* The intellectual doesn't like the idea of the machine doing it better than he does, but it's no different from the guy who was surpassed physically. Intellectuals are threatened, but they needn't be—we should only worry about what we can do ourselves. The mere idea that we have to be best in the universe is kind of far-fetched. The fact is, I think *we'll be enormously happier once our niche has limits to it.*[43]

Why should we worry about whether thinking machines will someday surpass us? Lifting machines already have. But who wants to lift anything? Only the intending human subject, still agreed to turn the machine on or off. Fredkin applauds the assistance given by machines in the tempering of human hubris, in the revelation of limits. But much simpler machines have already helped us fix our precise place in the expansion and contraction of universal scale, by showing that our reach is governed by our capacity, by providing a stream of analogies to explain ourselves and our worlds. If the computer can be said to fundamentally improve upon us, it would only show that a part of our reason is better than the whole. Pride in logic is increased as faith in humanity goes down. The limit clarified by the computer is one within humanity, not outside us. And yet the wall that logic finds is one that we often pass through whenever we need a part of the world which does lie beyond our capacity to divide and understand.

The image of ourselves as universal machine seems more believable as computers improve. We may not be reducible to the technology of today, but if the approach is sound, we can only get closer. What are the consequences of the spread of the belief that humans are merely superior computers, just one step beyond what we know how to build? Sherry Turkle's *The Second Self,* a sociology of computer users and a study of the way the computational metaphor has crept into our culture, suggests that the effects of this belief are quite diverse. Generalizations about the advance or danger in considering ourselves as vast, ultra-flexible computers are bound to be as naive as any other blanket statements about humanity. A patient lies on the psychiatrist's couch and says "I seem to be a machine" to reflect his despair over an inhuman, impersonal, and regimented life. A woman at a party of computer science students grows impatient over the debate: "I don't see what the problem is—I'm a machine and I think," believing herself to be a "collection of programs" that only produce a unified self through their interaction.[44] One person's prison is another's promise.

Those who identify the metaphor of the machine with all that denies humanity have forgotten how deeply technology has been entrenched in all our attempts to know the world. On the other hand, some technologies demand much more from us than others. The computer does not need constant supervision while following its programs, and in its detachment, its power as metaphor seems to grow. Instead of suggesting a picture for the way we should relate to the world, it provides a description of rules which may be applied to our minds *or* the world with equal ease. Upon this similarity, there are some who say that the computer is a working model of both thought and nature which at last reveals their connection, decentralizing our conception away from the self and towards a universally applicable set of laws which has been proven to work. If we choose to believe that the mind is like a multiple data processor running many programs simultaneously, we do not need the idea of an organizing ego to hold the stream of calculation together. The adherents of this view will not try to explain what cannot be explained, calling these moments the coincidental intersections of different information paths.

Some applaud this direction in thinking as a transcendence of anthropocentrism, moving at last beyond the limitations of a view of the world as always experienced by a thinking subject or perceiving self. If everything can be mirrored as a program, are humanity and nature alike finally united upon a single grid of unassailable logic? Once again, the idea is contagious. Some people who begin to program computers imagine their mind to be divided into those processes which may be imitated by the machine, the logical part, and those which elude description—the animal or instinctual. Yet after a few months on the machine, becoming impressed with what the patterns on the abstract device can accomplish, they begin to want to find a way to model all human (and other worldly) processes by streams of calculation. Technological methaphors are so infectious they are based on things that work.

For example, consider Turkle's account of the changes in the conceptions of mind of Arlene, a computer-science major at MIT who is also an actress. At the outset of her first programming course, she has the following view of her own mind:

> There is a computational part, that's the part that does . . . my
> reasoning, my logic, my math homework. . . . But then I have
> *another system*. It is built up from instincts. Evolution. My ani-
> mal part. It is involved with love, feelings, relating to people.
> It can't control the computer part. But it lives with it—some-
> times fights with it. And this is the part that gives me the feeling
> of being me.[45]

As a fresh, first-time visitor to the world of the computer, Arlene recognizes that what she has learned of logic will be useful in the new land, but that an important side of her experience will resist translation to the foreign tongue. However, three months into the class, her views have been modified by a firsthand initiation into the ways of computers. Her own mind now seems susceptible to the same analogical treatment:

> In general I see my mind in terms of continual processing by
> internal programs. But the weight given to the output of these
> programs can be influenced by emotion. And then when they
> come up to consciousness, they come up to a level where there
> is this other kind of agent—the *special agent*. The one in
> touch with my history and with evolution.[46]

Emotions and the animal side of our lives prove a challenge to computerization, but someone firmly immersed in machine logic will want to look therein for an explanation. History is no longer part of an untouchable realm for Arlene, but is instead apprehendable as a "special agent"—mysterious, but an agent nonetheless, something with which the computer could conceivably deal if instructed correctly.

The word *self* may be bypassed in a model that finds a counterpart to every function of the mind in the virtual world of the computer. If it works, all we know of ourselves and the world can appear under the umbrella of a single pattern, or at least seem plausible as a construction built out of very basic rules. This would unify humanity *and* nature around digital instructions. And when the imitation replaces the initial apprehending in our minds, one particular kind of thought has revised the rest of experience enough to claim to include it within a world assembled out of Turing's laws.

Roger Penrose reminds us that however powerful the deceptively simple rules of computational mathematics may be, they still do not encompass the elegance and beauty of the kind of absolute algebra and geometry which inspired Plato, Kant, and countless others with their claims of unequivocal clarity.[47] The computer repeats procedures with tremendous speed, successively approximating the answer. The absolutes of nothing, infinity, and perfection are never subsumed by the endless series of zeros and ones. Perfect shapes and numerical facts are still not encompassed by the calculations that edge towards them—never mind ecstasy and grief, as even the more sublime reaches of the generic machine's familiar mathematical turf are only mirrored, never included. We should not forget that it is the *distance* between ourselves and certainty which has inspired so much searching in philosophy. If we bridge this abyss with logic proposition by proposition, some part of the eternal is lost to us, even as we approach it.

VIEWING THE VIRTUAL

This model world of humanity and nature unified by the logical engine of the computer may be extended far, but it also hides things from us as it expands. Machines change our experience in a manner far more profound than analogy. As technologies extend our capacities and sense of self, our world is enriched. But technical success does not cancel the significance of intention.

Such an attitude is heretical to the original dogmas of the first computer developers, who envisioned that their futuristic machines would one day take the place of human scientists and engineers so that we would have more time for other pursuits. Yet the reason computers have become so ubiquitous and accepted today is that they have evolved into more modest devices that can more clearly extend the capacities of many different kinds of human workers. Less independent, they are more useful when *tied* to us. Word processing is accepted and understood by far more people than those who choose to play Turing's games, and computers are used in arts and design almost as much as in telecommunications and business. Such machines are called "user-friendly" if they allow easy interaction with their intending users, who do not ask to be fooled into thinking their computer is intelligent, but want it to be a sophisticated *tool,* which should become as seamless an extension of their own creative thinking as the pencil is a refinement of the movements of the hand.

Doug Engelbart is credited with being the first to see the computer's potential as a tool for all human writers and thinkers, not just those scientists and codebreakers who needed to crunch a lot of numbers fast. He imagined the universal machine as something which would fulfill its potential by "augmenting human intellect" rather than substituting for it. The future that he envisaged in 1963 was one in which "hunches, cut-and-dry intangibles, and the human 'feel for a situation' usefully coexist with streamlined terminology and notation, sophisticated methods, and high-powered electronic aids."[48] A computer which could be a part of this future is not necessarily one that thinks like a human, or a machine which encourages a person to think like it. Instead, it would need to be a device which could help fulfill human purposes, and be easy to engage with so that ideas, not just quantities, will be able to flow.

In addition to conceiving of the first primitive word processor, Engelbart invented the hand-held "mouse," which is at present (1993) the most common method for a user to indicate instantly a precise location on a computer screen, either for drawing pictures or accessing graphically presented data. He connected systems that enabled individuals at different terminals to converse with each other through the link of the screen, creating the new kind of meeting known today as computer conferencing. None of these applications was thought of by the early computer pioneers, because they had forgotten that even this newest of machines is first and foremost a tool, still best and most complete as an extension of human abilities. Though most were initially unimpressed by Engelbart's ideas, by the mid-sixties the whole field of computing was switching over to an interactive model,[49] where computers are not primarily to challenge people, but to assist us.

This redirection was necessary to change the image of the computer from a terrible, frightening, superhuman device into a friendly fact of life. Twenty years ago many were afraid that computers would soon surpass them at whatever they could do, stealing their jobs. Today more and more people find a way into the world that the computer creates, through input and output devices that encourage a more flexible and direct manipulation of digitized material by human hands. The computer now begins to alter our thoughts with the flexibility it bestows on whatever material we are able to store in its banks.

This liberation is not neutral. It requires more from the user, not less. All of us who have used word processors have felt the tremendous freedom that comes with the writing of a text that can be easily reshaped and rephrased, without worrying about disrupting the sacred order of the typed or penned manuscript. Making changes is simpler than ever before. Moving whole paragraphs and chapters around each other is as easy as striking out a single character. The word processor gives us great freedom of choice, but it does not tell us what to say. It seems almost entirely opposed to the idea of the intelligent machine, which might do the editing for us. But the two are not as distant from each other as they may seem: both are designed to fulfill human wants. One challenges us to assess what we know of our minds by facing us with our own logic in games or in the solving of problems, and the other opens up to an infinite, empty, and virtual world which we are challenged to fill.

The world conveyed by the interactive computer has been dubbed "virtual" because its location or features cannot be pinpointed in the tangible world. It exists within the relation between the machine and the user. We cannot place it inside the machine, because it is not there unless we invoke it, and it is not wholly within our minds because we do not possess the hardware necessary to conjure it up. A virtual world might also be said to exist between the reader and the book, but we are powerless to alter the shape or form of the book as we flip through its pages. In the computer, all parameters can be changed, and we can move throughout a constructed universe of our own making, on virtual paths invisible even as we tread upon them.

A program that helps us generate texts is only the simplest example of a route into such a world. A drawing and painting program that lets us change images on the screen *as if* we were directly touching them is slightly more advanced. But both these examples are hampered by the fact that we usually expect a printed, fixed output from either of them. Writers are usually writing books with their word processors, not flexible texts which they will modem out to potential readers. We are used to flipping pages, and the computer helps us print things that look like familiar pages. Illustrators rendering graphics on a screen usually expect their work to end up on a printed page. The virtual techniques are in the service of a tangible end. The computer is just a stage which the idea passes through in its route from mind to paper. (This will change if the screens improve enough so that information is easily read from them. Some people already feel the improvement is there, but most still find books more flexible and easier to access.)

"Virtual reality" is the phrase used to describe the world that only exists along the path of extension from user to computer. It is intensified through more sophisticated means of interacting with the machine, such as data gloves and data suits that track the motions of the hand and then even the entire body. Special earphones and eyephones would allow us to see and hear the constructed world; of course, ways of stimulating nerves throughout the body would really be necessary to complete the circle of virtual interaction, to make us more than instigator and observer. Such things already exist in primitive form, but they are a bit like playing with television in the 1930s; the idea is so far much more exciting than the reality.

Why would anyone want to enter such a world entirely of their own making, without the surprise of tangibility? Jaron Lanier, one of the inventors of the data glove and a pioneer with high hopes for a virtual future, imagines that such a flexible form of extension will bring humans closer together upon grounds previously far too abstract to share:

> What's remarkably beautiful is that you can make up [the world] in virtual reality and share it with other people. It's like having *a collaborative lucid dream*. It's like having shared hallucinations, except you can compose them like works of art, . . . in any way at all as an act of communication.[50]

Once again the future of computing seems far-fetched and all-promising. Virtual reality is much more exciting in science fiction, where it was dubbed "cyberspace" by William Gibson, who has since written that he is not so interested in practical developments in the technology because it is all the more interesting to imagine what it might be like through the freedom of the pen.[51]

Most of the more interesting speculation about virtual reality is more fiction than science. Donna Haraway's feminist explorations of the idea of the cyborg as a futuristic body-machine more friendly to the feminine than previous machines is more optimistic cloak than careful response to existing technology. "I'd rather be a cyborg than a goddess," she writes, without being very sure of what a cyborg might be.[52] It is much easier to mythologize technology while dreaming about it, not trying it out. This is a bit like the expectations placed on robots in the public imagination of the mid-twentieth century. Now there are robots all around us, performing many rudimentary tasks as a part of automated production. But they are not so exciting to contemplate, because they are not very much like human beings. They are still extensions of human beings, and only make sense when we use them to realize and change our ends.

The information world may not now promise an intelligent companion, but instead the ultimately free transmission and reception of our most intimate secrets, if properly formatted and digested. As poet Delmore Schwartz wrote, "in dreams begin responsibilities." What kind of dreams will we want to share with each other over the network of imagination? I am not certain that more can be conveyed in a digital, fluid realm than through set expressions in word, image, or sound created by more traditional means. But machines that can connect to human faculties hitherto untapped will certainly change at least the way we consider our untempered experiences. Lanier himself is reflective enough to realize his creations do not disavow any other reality, but do affect our appreciation of the world apart from machinery:

> I tell you, the most vivid experience of virtual reality is the experience of *leaving it*. Because after having been in the reality that is manmade, with all the limitations and relative lack of mystery inherent in that, *to behold nature is directly beholding Aphrodite*; . . . a beauty that's intense in a way that just could never have been perceived before we had something to compare reality to.[53]

So digital extension is most poignant when we may turn it off! This may be the most telling comment on the future of computed worlds. We will only accept the course of disembodiment if we are able to turn our back on it at will, returning to a more grounded and direct alternative. This is something like the reading of history which reveals that no one chose to appreciate nature as some wild sublimity detached from humanity until our civilization was built up enough to appear as competitor to the rest of the world. Reality emerges so vividly as a nature greater than what we can ever know *only* if our attempts to know seem somewhat convincing—assuming that nature is a place out there which remains consistent. Yet this chapter has shown that its qualities change so drastically with the progress of technologies that it may be impossible to retain any unified sense of "nature" apart from our continuously transforming attempts to learn it and to build a world to our own liking.

The otherworldly promise of a universe accessible only by computer leads the virtual voyageur back on a course to a nature more alive than ever before. We cannot escape this paradox: even as technology clearly changes nature through our moves to explain with it, it still may be justified as an excursion that rekindles appreciation of the "real" world. Coming back from virtual reality sounds like coming down from an artificial high. We think we have learned something: "Isn't sensible reality so much more pungent and rich than whatever we try to replace it with?" But we only ever succeed in circling around this nature, redesigning it with the aid of our evolving tools.

The concept of virtuality describes a kind of interactive computing far removed from the way in which we usually use tools, as implements to help us complete a specific and bounded action. The promise of a whole world tying us to the black box suggests an encompassing video game which we actually inhabit. We all know we cannot live inside a virtual world, because it just takes the body as a source of stimulus and a destination for response, once more the limited system closed by input and output. The body is never a fact of this new world, only a node for connections. We may never live as complete human beings in the life space of the machine, but many of us already work there. We spend hours engaged with the screen, manipulating symbols through the keyboard and finding ideas in the reach between hand and eye, bound by complicated circuits we don't need to understand as long as the tool functions smoothly and transparently. Yet it is never transparent if the location of our work is found in our engagement with the computer. What I am now writing does not exist anywhere else but in my bond with the machine: until I "save" it, no one else could extract it from any disk, soft or hard. At the end of the day I will have wrestled with sentence structure, paragraph formatting, subtitle choice and placement, planning for the next chapter, all within the mind extended through the device. I and many others already work in virtual reality, even if we might not want to admit it is a real place.

And when I look up from the screen and listen to the wind from outside the window, rushing into the room, slamming doors and blowing papers off the walls, I know there are forces that offer more surprise than a sudden crash caused by an unsuspected virus in the hard disk. All this may seem obvious or trite, but the point is to show how easily we get used to spending our entire working day in a world that offers no sense of place, one which is created out of inert circuits the moment we invoke it. Sure I laugh at the wind when it makes a mess of the desk, but the emotions are less certain when I consider it may no longer be the same wind.

A computer programmed to be a sound synthesizer enables one to create an immense multitude of sounds, without offering criteria by which to judge one over another. We are first freed by considering sounds purely as themselves, until we get up from the machine and its virtual tones. Listen to the wind right after inhabiting the world of virtual sound: the first thought is how its voice could be emulated on the machine. I imagine how many oscillators are at work, what kind of filters are sweeping across the frequencies. While appreciating the unpredictability of the wind's music, I still try to categorize it in terms of what I am able to create.

These vignettes are intended to make the arena of virtuality more immediate and experiential in the present age, not a future one. The virtual is really with us whenever we work in a world of our own making. The change it brings to the lived world does address a certain limit in the application of human thought. It asks: how much can the disembodied mind know and do? This is not a question of too much thinking and not enough doing and feeling, but of the assessment of an entire direction in human development. Because the computer acts upon fixed rules and we so far do not, the attainment of any ends upon generic machines only occurs when a great distance is crossed from us to them. The computer is an extension, to be sure, but its manner of operation is initially quite far away from ours. We need to think at least in parallel with it if we are to make it work for us. And when we learn to be like it we are suddenly very close: nature almost seems to follow a plan that easily encompasses us as well. But the gap is still very real. We do not become the machine, even when it seems invisible.

The computer is a universal and generic machine because its mechanism may be detached from any particular end we wish to realize with it. First it was understood as the embodiment of regimented thought, theoretically capable of following any task which we might wholly explain to it. Now it looks more like a surface on which to play out elusive dreams so that we may share secrets with each other without needing to touch. It promises whatever we are able to meticulously describe according to codes detached from physical experience. It tests the mind as well as the human spirit. First, how much can we accurately describe and predict? Second, how far away from physicality can we venture and still accept as experience? The virtual may be melded with the real if we can be said to feel at home there.

The computer was once touted as a challenge to our sense of ourselves; it might fool us into thinking it was intelligent, and therefore similar enough to be mistaken for one of us. As we have advanced it, the promise has changed into one that coaxes us in, with the offer of a freely determinable world, not hampered by the mundane strictures of physicality. The limit to what can be accomplished in virtuality is the same as the limit to what the mind can accomplish without the life of the body. This is how the computer makes nature as examined stand for nature as a whole; the power of the machine increases when we imagine it to be more than an intelligence, now a whole world. This world is a place we can enter and enjoy.

What we shape in the virtual world need not be bound by the laws of physics. Yet those most excited about the potential of the intangible are thrilled that we will be able to feel like we are in there, that the sensors on our body suits will convince us that we are directly shaping the unbounded sphere of possible events inside the machine universe. They are hoping for contact, the touch that leads to a direct empathy with the objects of our creation. They want a tool more like the potter's wheel, less like the engine turned on with the flick of a switch.

The feel of smoothed objects shaped by spinning, the predictability of a world that recognizes the clock, the independence and self-regulation of the engine, and now the ultimate challenge of a technology that quantifies the abstract—these are more than sequential phases of innovation which reconfigure our concept of nature and humanity together. Today they appear as concurrent facets of a mind and a universe which we want to believe *works*. We want to touch it, to imagine that it possesses order, and to expect that it somehow rules itself. And we wish that it prefers the simple to the complex.

Earlier machines worked on our minds as somewhat of a side effect to their intended purposes of making vessels or powering ships or trains. The clock imposed an abstract, unvarying rule on the fluidity of time, but it did not do any physical work. The engine gave us a running source of power, but the analogy of considering organisms as self-regulating systems was a step back from the practice of the tool. The computer, in contrast, demonstrates *the pragmatic promise of abstraction*, so much so that its users are finally tempted to leave the physical world behind. By this point, the transformation of thought becomes central and no longer ancillary to the use of technology. Computer technology's stated purpose is to change the way we think. The reach of the mind seems to lift and to soar, as we learn to imitate and simulate more and more of the surrounding world. And if even the chaotic can be quantified, the whole notion of order is extended beyond its formerly regimented, unbreachable boundaries. We cross the borders while forgetting why they are there.

But mind does not live if separated from body. *Under the ground we hope there are figurative wheels and pistons too, churning against each other, justifying any vision of system. We need this predictability to draw water up the well. "Either everything is a machine, or it is nothing at all."* Or does only science demand this? Do not forget that the main stream of technology does *not* extend the mind, but expands the range of human physical force. And although we seem to encompass still wider areas with our attempts at understanding, we are also able today to destroy much more than we are able to comprehend, both intentionally and unintentionally. The effect of this capacity on our sense of place demands more than any measure of silence.

Ramelli, as a military engineer, drew many siege and assault devices among his Various and Ingenious Machines. There are catapults, door removers, stealthy bar cutters, and expandable bridges like the one shown here. The walls of the enemy city have been shattered, and the walkway is flexed across the torrent to let the army enter with ease. The troop with the technical advantage is extended right into enemy territory, and the path can be pulled back at will.

Military supremacy has always been used as an argument for the pursuit of technological advance. Ramelli knew this as well as anyone, and probably was aware that it was in the service of the military that the most fantastic of his creations might conceivably excite enough interest to encourage actual construction. Reports of the value of these inventions in real battles have survived, though all of the structures themselves have returned to dust. Still, these byzantine implements of warfare seem quaint compared to the weapons unleashed in our present century, soon nearing its close.

The computer may have freed our minds to build entire worlds above the confining facts of nature, but its abstract power still pales beside the transformations wrought by those technologies which extend our hands to alter the world in the fulfillment of our own, sometimes cruel plans. We can destroy things long before we understand them.

Two kinds of possible destruction cast an unmistakable pallor over the present age. The first has been inflicted only twice, with the atomic bomb explosions in Hiroshima and Nagasaki, but its promise of total destruction continues to haunt us, having already changed what it means to be human by bestowing the capacity to completely denude our environment in any war that escalates out of control. Pushing weapons technology to this extended limit means either the end of our species or the curtailment of war. Either way, we are transformed by the idea of the atomic bomb, a machine far more elaborate than any Ramelli imagined. We possess the technology to destroy ourselves and our world, so the Earth no longer seems eternal. We have a much harder time imagining immortality for ourselves or for the range of nature. The limitation of this particularly human tendency to kill becomes clear.

The second kind of destructive limit is not the outcome of any simple intention like the strengthening of one's defenses against the enemy. We are also transforming the tenacity of our environment through the collective desire to make the world more amenable to human interest. Industrialization and domestication of the planet brings with it pollution and ecological upset that may warm the temperature of the planet so much that it might no longer offer the promise of hospitality to the species that worked so hard to tame it. We only want to make the Earth a safer place for us, but in the narrowness of human self-interest, we forget how much we depend on concerns and currents other than our own. No one intended the greenhouse effect or the hole in the ozone layer, but these changes are the direct result of the accumulation of technologies designed with other, more "noble" ends of human betterment in mind.

Both the warming of the atmosphere and the ability to let loose atomic disaster are explosions in the history of humanity challenging the resilience of nature. This is a nature that emerges as final arbiter of human arrogance, however it may have been modeled with images sifted from the history of invention. The Bomb may have been conceived as necessary military strategy, but it demands control of our destructive urges. The greenhouse effect suggests that the full weight of all human technology may be inadvertently destructive, leaving us even less of a recourse. The extremity of these two directions shocks us into finally asking what is beneficial in technology and what is adverse. Once admitting the real likelihood of disaster, we need to find a way for individual techniques to continue even with total danger so close.

The extremes presented in this chapter are important not because they make good examples for the behavior of technical extension, but almost for the opposite reason. Tools that question the very validity of human habitation reveal an uncertainty inherent in the intention behind even the most mundane techniques: the furtherance of a single desire unaware of context. Extreme cases make bad examples, though they are the strongest ones.

TOOLS UNLOVELY TO SEE

Destruction is an intent which may be realized by any of the three successively detached modes of extension of human action in chapter 2. Yet an individual's physical strength is nothing compared to the damage we might want to cause. Eradication of the enemy can now be enacted by a single telephone call, and the pressing of several buttons. At least that's what they tell us. Death becomes a diffident urge, accessible by remote control. The outcome need never be confronted by sight or by touch.

No country has launched a nuclear warhead in combat since World War II. The existence of weapons of such tremendous destructive power has forced us to reconsider war as more than a game in which we should do whatever it takes to win. With bombs this dangerous, war must be closely watched so that it never escalates into a battle which could be stopped only by total destruction of all sides. Some think mutual destruction by vying superpowers will be the inevitable and final war. Others argue that limited nuclear war might still be possible, remaining a scenario we should be prepared to fight and win.

The particulars are not at issue here. Whether or not their fear is wholly justified, many people now *believe* that we possess the power to end the world with the sudden depression of a trigger. It was the Bomb of the 1960s that first offered this danger, but the sentiment existed already at the close of World War II. And we do not all trust in those who dutifully wield the trigger to have full grasp of the implications of the machines they command. On the other hand, if the greenhouse effect is a real and inevitable consequence of the more well-intentioned changes we have inflicted on the world, destruction of nature may be the unstated end of our culture. We may be heading for it whether we try to stop it or not.

Life bounded by these concerns soon leads to a desire to question technology, to wonder whether it has not been fundamentally destructive since its beginning: are these its inevitable limits, accidental or planned? The previous chapters have considered how technology, though striving towards an ideal nature outside its bounds, ends up reconstituting that nature through its offer of increased explanation and power. Now it may be able to destroy even those things which it is unable to help us know. This is the tragic upper hand of destructive prowess. But the returning arc of its influence requires us to reconsider our plight.

The aftermath of the atomic bomb encouraged popular disenchantment with technology in the years following the nuclear conclusion to the war in the Pacific. Many wondered if the newly assumed ability to destroy life on Earth was the culmination of a technological orientation that had been negative from the outset. (On the other hand, prosperity and automation pressed on. This was still a minority position.) Humanity has always required technology to dominate and to know the extent of the world. Does its fruition in a tool of ultimate destruction prove that the drive to make the world more our own is fundamentally misguided?

To condemn all technology is to negate human history. But to refrain from being changed by the existence of the Bomb is to deny an aspect of danger which may always have been with us. If we follow Karl Jaspers, this risk was present at the very outset of human evolution:

> When humanity embarked on its yet-to-be-created history, we unwittingly risked our lives in a relatively brief span of a few thousand years. *Necessity and evil*—in whose conquest we rise above ourselves—*drive us on or devour us*. In failure, we would not live on like the animals but would destroy ourselves and all life with us.[1]

This ceaseless peril reveals itself as a contrast: technical intentions have been extended with the refinement of machinery, while it is still difficult to articulate any solid morality behind those wars which continue to be fought. Now the ultimate weapon challenges the very intention to kill an enemy. The promise of escalation demands a careful regulation of warfare. This kind of thinking calls the whole intention of one nation fighting another into question. The desire to kill has been extended to its final end, but what of the need to live as a community of people at peace? The implicate horror of the intention to destroy now continually haunts us. The only way to survive is to articulate ways in which differences may be tolerated and resolved without the frenzy of war.

And yet wars have so fueled human development in the past, both in the limited sense of resolving economic depressions and in the impetus they have given to innovation in technology and science. Battles were traditionally fought for a stated and accepted reason. Before the bomb, it was at times possible to believe in them, and to participate in the boon they offered to national progress. But mine is the first American generation that has grown up without knowing what it was like to live through a war which the nation wholeheartedly believed in. (The Iraq-Kuwait skirmish of 1991, blown totally out of proportion by the media and the presidency, does not count.) A whole wall of experience seems to stand between my lifespan and that of those who lived through World War II. What remains of that time for those of us who have only heard or read about it are the stories of the mass death of the Holocaust and the terrible technique of the atomic bomb. At times both seem almost too extreme and cruel to have actually occurred. Is it naive to believe we live in a "kinder, gentler" time today? Of course.

We may not have been actively involved in mass destruction, but its undeniable history has defined the stand which we take towards our durability in the universe. Though the computer has crept systematically into the way we think about all things, the probability of nuclear destruction has become an unshakable context within which we find either despair or hope. Despair comes if we resign by imagining there is nothing that can be done about the inevitable end; hope arrives when we consider that it is the knowledge of this weapon that changes us, not its use. Realizing the end which unassailed destruction will bring, we want war no longer to be possible. Instead, we wish for a positive sense of peace which is not devoid of conflict, but a kind of harmony among peoples which can still build upon innate differences.

Yet peace has traditionally been transitory, a pause between contests that enthrall and define a nation. Recall Machiavelli's advice to Lorenzo de Medici: he says the only *techne* worthy of serious consideration by the Prince is the art of war. For centuries, this was the prime way in which empires were built and fortified:

> [The Prince] ought, therefore, *never* to lift his thought from the exercise of war, and he ought to exercise more in peace than in war, which he can do in two modes: one with works [*opere*], and the other with intellect [*mente*]. And, as for works, besides keeping his own arms well-ordered and exercised, he ought to be always out on the chase, and by that means to accustom his body to hardships and also in part *to learn the nature* of sites: to know how the mountains rise, how valleys open, how the plains lie, and to understand the nature of rivers and marshes and in this to put the greatest of care. This knowledge is useful in two modes: first, he learns to *know* his own country, and he can better understand how to *defend* it.[2]

This prophetic passage outlines the historical link between knowledge and defense and illustrates how war has been instrumental to the development of civilization and culture. Nations unify themselves when poised against the enemy, and here even natural science is motivated by a desire to comprehend enough of the lay of one's land so that it can be preserved against attack. Here is a vision of science as part of politics, defined such that a government whose prime goal is preservation of the nation should deem to care about the furtherance of basic knowledge. The argument continues to be used to fund research unto the present day.

And certainly the urge to curtail the deployment of nuclear weapons is a fight with a part of ourselves: the one that is vastly inspired by the intention to destroy in the name of ourselves or our countries. To hold back this "discipline" that outpaced all others in the concerns of Machiavelli's well-behaved Prince, take heed of a much earlier account of the strategems of life and war:

> The world is won by *refraining*.
> How do I know this is so?
> By this . . . :
>
> Weapons are tools of bad omen,
> By gentlemen not to be used;
> But when it cannot be avoided,
> Then use them with calm and restraint.
> Even in victory's hour
> *These tools are unlovely to see*;
> For those who admire them truly
> Are men who in murder delight.[3]

In these words of Lao Tzu, the greater the power of implements of death, the less lovely they seem. Today war is possible only with restraint; if we want to kill as much as we can, all may be killed, friend or foe. This is the promise of the "best" weapons, the most perfect machines of death. Now we cannot even use them. Thus the entire intention to wage war with the single end of victory for one side over the other is called into question.

Since Hiroshima all wars have been limited wars. And we have been more reluctant to embrace them, knowing where the intention to fight may well lead. Nuclear-capable superpowers often deny their clandestine involvement in third-world conflicts. War is more than an embarrassment, as we have within recent memory the indiscriminate image of mass killing that comes with the detachment of a weapon of unlimited destructive power from our limited intention to win. Here is the first intention which technology has threatened to *nullify* with its success. Here is a machine we hope we will never have to make use of again.

Its invention and demonstration lead some toward philosophy. We want to articulate how it is possible for us to survive after knowing the terror of our own hands. We ask if knowledge may be independent from the uses to which it is put, preparing to be responsible for any of our actions within the circle of technological development, application, and reconsideration. This is the realization that challenged those scientists who worked on the Bomb—if they were not responsible for the power of what they had built, who was? The decisions which needed to be made concerned a tactical reality no one had foreseen, so nobody was prepared to deal with it. Evaluation of the morality of an ultimate weapon is not something anyone had been trained to address as a specialist. All who had done their small part had to accept the implications, which defied science's claim to be an independent activity tempered only by the limit of truth. In a world with the Bomb, we are all warriors, even if we resist its call to atomic arms.

RETICENT DESTROYERS

When science catastrophically alters the domain of its experiments, it can no longer be an unswerving investigation pure in its commitment to the revelation of nature. Two reactions to the Trinity test at Alamogordo illustrate the opposing directions the minds of onlooking scientists were extended following the first successful atomic explosion out there in the New Mexico desert. J. Robert Oppenheimer is often remembered for quoting the *Bhagavad-Gita*: "I am become Death, the destroyer of worlds."[4] With its overwhelming power, the Bomb requires us to inhabit fatality itself, irrespective of whatever life it is used to cut short. This is human power extended into a chilling detachment from life, an explosion that immediately evokes the end of the world. This moment may have been the beginning of Oppenheimer's disenchantment with the project of nuclear proliferation. With the shock of the explosion, he certainly grasped the awesome danger. If humanity becomes death, what will we be able to recognize of life, save its end?

On the other hand, I. I. Rabi, previously afraid of committing himself to a physics that culminated in a weapon of mass destruction, seemed to perceive a new mastery of natural force in the test explosion, more enhancing than the doubt that comes with the inhabitation of death:

> It was a vision which was seen with more than the eye. It seemed to last forever. . . . Finally it was over, diminishing, and we looked toward the place where the bomb had been; there was an enormous ball of fire which grew and grew and it rolled as it grew; it went up into the air, in yellow flashes and into scarlet and green. It looked menacing. It seemed to come toward one. A new thing had just been born; a new *control*; a new understanding of man, which man had acquired *over* nature.[5]

Rabi thinks that we now have something *over* nature, some kind of one-upsmanship of command. He seems more ready to embrace the event as a gateway to new knowledge, rather than accept the toll of responsibility which Oppenheimer's grasp of destruction would suggest. But does the explosion itself offer the promise of this new understanding, or does it rather demand more from us, without so much as hinting at the way to find it?

Gaining the power to release death without discrimination can certainly be called a new understanding. The question is: does it constitute any further *mastery* of nature? Like many technological inventions, the Bomb lets us grab hold of a previously untappable force, extracting this force into prominence. We take the exploding atoms out of nature, and we lose their context. This explosion triggered by humanity does not include the promise of life. It is not bounded by any cycle that could temper its effect and include our continued survival.

The Bomb is probably the first technology that pushes a recognized human intention to a final limit: beginning with our perceived need to destroy an enemy, we have extended our reach to be able to destroy the world. And no one wanted to intend that result. Yet the "success" of the bomb illustrates how much easier it is to extend an intention than to temper its effect.

Becoming death does not so much imply a new understanding as require one. Before the War, Einstein wrote the letter alerting the American government that an atomic bomb was possible. After the War, he wrote that the atom bomb had changed "everything save our modes of thinking."[6] The new kind of thought needs to be one in which Machiavelli is no longer the guide—today's rulers cannot be obsessed with war in the manner of his Prince. Cooperation at the international level becomes far more important than the shout of one nation's interest against another's. For now we are all afraid of the final end of any untempered war.

It may seem a paradox that the ultimate weapon can never be used, and that any currently conceivable conflict needs to be smaller than wars which have already been lived through. Wars are by no means over, but they are all cautioned by the memory of the moment in which, though gaining control of certain forces in our environment, we lost control of our reasons for needing this control. Not only scientists recognized this transformation. Henry Stimson, as Secretary of War, initially lobbied strongly for the authorization of the Manhattan Project, as it was his job to make us win the war. But when he realized that he had given the go ahead to a weapon that would kill women and children as well as military targets without being able to discriminate between them, he sensed that more than this one war was at stake. Fearing that the reputation of the United States as a nation founded on ideals of fair play and humanitarianism was in danger, he soon saw the bomb as an invention that would mark a historical epoch, nothing to be dropped for simple tactical reasons. His thoughts wandered far from the battlefield, as these handwritten notes before a committee meeting on May 31, 1945, indicate, resembling a tentative poetry:

> Its size and character
> We don't think it mere new weapon
> Revolutionary discovery of Relation of man to universe
> Great History Landmark like
> > Gravitation
> > Copernican theory
> But,
> Bids fair [to be] infinitely greater, in respect to its Effect
> > —on the ordinary affairs of our lives.
> *May destroy or perfect International Civilization*
> May [be] Frankenstein or means for World Peace.[7]

To ensure eradication or the perfection of peace—this Bomb seems to inspire the polar extremes of utopia or oblivion. But technology in its explosive culmination is neither the answer nor the problem. It is a fact of our time that changes us. What is most revolutionary about the harnessing of nuclear destructive power is not how we might use it, but how its very potential to reach devastation requires all of us to take a stand on it. It is the moment when people learn to kill more than they can understand.

Physics is supposed to be the science that investigates the mechanical building blocks of the natural universe. As such it is inseparable from a faith in a universal mechanism. Whether Newtonian or relativistic, it still believes there are fundamental rules which we may discover. But science is never separate from technology, in either its means of extracting data or its appropriation of technical metaphors to explain its discoveries. We need machines before we can imagine nature to be mechanical, and when science lets us change nature in the direction of our disposition, it most of all proves the power of the model of mechanistic rules. Too bad these rules promise devastation before they explain what they can destroy.

And no one is sure whose responsibility it is to keep watch on this use. This is as true today as it was in 1945, and we have not advanced far in our willingness to take a stand on the inventions we create or apply. The assessment of technology is not only evaluation of the economic value of a particular tool, but a consistent awareness of what the tool is doing to our place in the world. Oppenheimer, reflecting just after the war, says scientists should pledge allegiance to the search for truth, and this view always includes considering the consequences of the application of one's discoveries. The scientist must believe that whatever he discovers is of value to humanity, and must accept the task of locating this value:

> When you come right down to it the reason that we did this job is because it was an *organic* necessity. If you are a scientist you cannot stop such a thing. If you are a scientist you believe . . . it is good to turn over to humankind at large the greatest possible power to control the world and to deal with it according to *its* lights and *its* values. . . . It is not possible to be a scientist unless you believe that the knowledge of the world, and the power which this gives, is a thing which is of *intrinsic* value to humanity, and that you are using it to help in the spread of knowledge, *and are willing to take the consequences.*[8]

Here is a picture of scientists "organically" trapped in the march of progress, yet responsible for whatever position they hold. What is curious is that the scientist appears helpless and unable to turn away from a potential discovery (if they don't do it, someone else will), but then it is imperative that they stand up for what has been discovered. World-shattering discoveries thus lead science closer toward politics, and scientists are forced to think about things which formerly could have been left to others. For these "others" will not be prepared to think about the implications of total destruction until they have learned more science. *Everyone* has to learn more to gain any kind of clear grasp of the situation.

Edward Teller, another physicist whose career was charted at Los Alamos, has a different view. He thinks scientists should make pronouncements only in the areas of their expertise. Beyond that, they should keep quiet:

> The important thing in any science is *to do the things that can be done.* Scientists naturally have a right and a duty to have opinions. But their science gives them no special insight into public affairs. There is a time for scientists and movie stars and people who have flown the Atlantic to *restrain* their opinions lest they be taken more seriously than they should be.[9]

Ought all things which can be done be done? If scientists do not consider which things *should* be, who will? No one involved in an invention that so transforms humanity can afford to shy away from consideration of its effects. Are we to believe that Teller's fifty-year crusade for bigger and better weapons is motivated entirely by a love of pure research and an insatiable curiosity about what kinds of bombs might be built? He escaped persecution by the Nazis in Hungary and then Germany, and once in the United States, he developed particular views about how the nation that took him in should defend itself. Teller's personal history of course affects the views which he continues to espouse, and he should realize that nuclear physics has led scientists much closer to politics than ever before. He is right in noting that physicists may have no special expertise in international relations, but it is up to them to learn what is necessary to evaluate the consequences of their inventions. No one else has this expertise, either.

But there are problems in this blending of expertise and naiveté. Mistrust ensues. We're not trained in each other's rules. Leo Szilard, another Hungarian emigré, was one physicist who suggested that a nuclear weapon was plausible, and urged the United States to begin constructing one lest some other nation get there first. Yet from the beginning, he grasped the political implications, and wanted more involvement in policy making rather than less. This led to his gradual exclusion from the secret encampment of Los Alamos, and he spent the wartime years occupied with political problems as well as those of bomb design, in isolation from the main group of scientists who were moved out to the desert. Constantly shadowed by the FBI, he traveled the country dabbling in politics, trying to convince the government that the new weapon would introduce more political problems than it would solve. Worried that the physicists did not know enough of Washington's policy in regard to the use of the Bomb, he was among the first to predict the arms race, a consequence which he felt no one knew enough to stop. Always an outsider, Szilard would never need to endure the humiliation which Oppenheimer suffered, when the director of Los Alamos eventually spoke out against the further proliferation of nuclear weaponry, and was forever ostracized from secret government discussions as his security clearance was revoked.

Physicists glimpsed the ensuing madness of an endless arms race, but most were not sure they knew enough to speak up. They were as reluctant as politicians to admit that they were transforming the nature of international community as much, if not more, than they were advancing science. Possibly the most courageous was Niels Bohr of Denmark, who realized upon arrival in Los Alamos in 1943 that "we are in a completely new situation which cannot be resolved by war." With this came a revelation of a complementarity in the meaning of the Bomb: along with its offer of total destruction comes an imperative for the world to unite around a discovered limit to human cruelty shared by all: "It appeared to me that the very necessity of a concerted effort to forestall such ominous threats to civilization would offer quite unique opportunities to bridge international divergencies." [10] It does not much matter who has the bomb, as its deployment entails a devastation that implicates all of humanity.

But what Szilard feared was truer to the outcome than what Bohr hoped for: the years following brought an escalation in nuclear weapons throughout the world, and it is only today, almost fifty years later, that the dismantling of weapons seems slightly more probable than their continued proliferation. Weapons were, at first, direct and dangerous implements, which made the thrust of a human hand more formidable and lethal to an opponent. Once, we could feel the direct horror of killing another human being with our hands through a knife or a sword. Then bows and arrows made a graceful human motion an intention in the service of death at a distance. We might kill while being invisible to our opponent. Here the detachment that makes war bearable begins. From there we moved on to a series of destructive devices which we need only touch on the trigger to set off. But the killing is still motivated by a specific human desire and act, aimed at a particular target. With a weapon of indiscriminate, outright destruction, we may intend the explosion, but we cannot survive its consequences.

This is a path of technological evolution which successively separates the intention from the result, eventually making the very urge to fight wars seem ridiculous and even implausible. Yet we continue to engage in them, we continue to intend death. There are ways to justify our luck of survival. *Still* no one is quite ready to take the responsibility for the limit that this technique has revealed. Is the urge to kill itself wrong? Or need it only be tempered by some outside constraint, some rule that defines the maximum allowable destruction? We still need to chart a course mediated by despair and hope.

BETWEEN UTOPIA AND OBLIVION

> We're living in the last few moments
> Of the last few days
> The last moments ticking away.
>
> We are living at the close of the day.
>
> Could it be the soul of me
> That says that soon it will be free,
> To look up in the sky and see
> There something—
>
> Oh I could feel it everywhere.
> I know there's something in the air.
>
> Could it be that this will be
> The day that starts eternity,
> The day that we've been waiting for so
>   long?
>
> Oh, I want to be ready.
> Oh, I want to be ready.
> Oh, I want to be ready.[11]

So sings Sister Royce Elms at the First United Pentecostal Church of Amarillo, Texas, home of the Pantex plant, where all nuclear weapons in the United States are shipped for final assembly as of the mid-1980s. A community whose working lives are so involved with the continued construction of devices of mass destruction is bound to come up with a unique way of justifying its current mission on the planet. If we are wrapped up in preparation for the end, we must likely be near the end. Their justification for life in the midst of this predicament is to dream of a wondrous entry into the hereafter—eternal and constant, the final resting place of all true believers. If our work is only a preparation for destruction, our faith must lie in a world beyond self-inflicted terror. We prepare for the final day. It is no longer a moment of judgment meted out by the whim of God, but will be an occasion plotted solely by humanity, now parlayed into the means for bringing the world to a close. Yet it is still only God who can provide solace that an eternity worth waiting for will continue after the outcome of our own foolishness.

This is the message of a millenarian religious faith that prophesies a heavenly time of wonder and bliss to succeed the Earthly abyss that will follow nuclear war. This belief is spurred on by the conviction that technology is ultimately destructive, that the human spirit may survive only if we trust in a Higher Power to preserve us after we have denuded our immanent habitat. This may be the easiest psychological crutch to counter the inner threat of nuclear holocaust, as it manages to find utopia *in* oblivion, with devastation the necessary means for our deserved rebirth in the most promised of lands.

It is understandable how such a belief finds followers in the town of Amarillo, among congregations so involved in the manufacture of lethal devices that would reduce much of the world to rubble if actually used. It seems reasonable to imagine survival in the form of miraculous redemption. Only those Americans far away from these factories of war can think of these hundreds of powerful weapons as symbols of strong defense, not meant to be deployed, built only to deter any necessity of use. If one builds a machine, one expects it to be used. Whoever lies close to the making of these instruments of war needs a strong faith in some mission that justifies not only their construction, but also their application.

The song of Sister Elms does seem to touch a chord of our era, resonating with the repeating echo of our own ingenuity—we have too much power, not enough restraint. Armageddon now appears more probable as human result than divine intervention. Some already glimpse the limit of our entire world, if not just one particularly intrusive culture or way of life. We may choose not to submit to this ultimate limit, but it is equally difficult to know how to resolve or transcend it. Human-induced millenialism is appealing because it allows us to be immortal, to share in eternity, even if we destroy our environment. There is redemption beyond the baseness of our physical imperfection and evil. It is a position which cannot really be argued with, since it depends so little on logic or proof. But it is also immensely sad, to anyone who trusts in the primacy or even the value of the life humanity has created for itself on this planet.

Another sense of immortality is deeply threatened by millenialism. This is the belief that, whatever happens today, humanity will persist; that our great works and ideas will still be able to be appreciated in the thousands of years to come. No individual may be immortal on Earth, but our collective achievements may last far into the future so that others may enjoy or at least consider them part of a common and human heritage. This is what enables a human person to feel part of a larger community that lives in more than the moment, inhabiting the past as well as the future.

Before any immense nuclear war ends this dream, the *image* of such an end takes hold of us. The blockage of solar radiation with excessive debris leads to the bleak vision of "nuclear winter," a cold, empty world where little life could survive. Scientists will continue to debate just how severe the climatic effects of a nuclear confrontation might be, but the image lies irrefutably inside us. But science is of little help here. The picture of complete annihilation through atomic explosion has been with us since World War II, heightened by the test detonations of hydrogen weapons several years later. It is hard not to be frightened by even the remote probability of this vision. Death, as psychiatrist Robert Jay Lifton eloquently points out, is *no longer* the tripartite challenge that should shape humanity, not end it:

> *First,* the knowledge that one dies; *second,* the . . . internal recreation of all that we perceive; and *third,* the creation of culture, which is by no means merely a vehicle for denying death, but is integral to the human cultural animal, and probably necessary for the development of the kind of brain we have come to possess. When the image of nuclear winter threatens our symbolization of immortality, then it threatens a level of psychic experience that defines our humanity. When it threatens our proximate level of experience, it undermines and instills with fear and doubt our sense of moving safely through ordinary steps in human life.[12]

Nuclear winter implies total demise, with no sense of cycle. Here is total emptiness without any humanity, nothing like the emptiness a Zen meditation is designed to enter, but the total denial of anything human.

Though the actuality of nuclear winter may be the ultimate tragedy, the *idea* of such devastation might be positively exploited. It forces us to consider the most terrible of technical limits, the possession of a technology which we should never have to use. Grasping its meaning asks more from us than any personal death. It demands us to conceive of a death encompassing our whole species, and then all life on the planet. This requires tremendous empathy, extended identification and a vast expansion of our sense of self. Taking hold of this penultimate horror becomes a necessary step to take toward the inhabitation of the unique reality of our current age. We need to apprehend its depth to insure that it will never come to pass.

This response, in contrast to the millenarian one, suggests that an answer may be glimpsed out of the *possibility* of oblivion. Of course, only faith in human discretion and caution can support this hope. And many do not trust their fellow humans this much. With good reason, if one considers the further military responses to nuclear attack that have been suggested, such as the Strategic Defense Initiative, which concocts a scheme of laser beams to deflect the enemy's aim. Such a purely technical response once more evades the issue:

> A technology of protection is now to counter a technology of destruction, and human beings become virtual bystanders in the cosmic confrontation. . . . Technology now emerges as not just the "shield" we are told it is to be but as a realm from which we can draw magical, life-enhancing power.[13]

This "Star Wars" answer to the public fear of nuclear winter promises a panacea of total protection. Like the Amarillo church's longing for the end, it is equally disturbing as a dream that the idea of the end can be deflected from the minds of the people through further and even more fantastic technology. But faith in defense wizardry ignores the important limit announced by the reality of the Bomb. Discovery of this limit begins to change humanity, and we need to let ourselves be changed. We must let the nature of war be transformed, and all share the ramifications of the discovery.

The threat of nuclear attack need not lead only to the intention to develop technologies of nuclear defense. If a single worldwide nuclear altercation contains even the slight possibility of destroying all humanity, the idea of it *may* bring everyone closer. Fear of nuclear war makes the world a smaller place, encouraging all of us to band together to protect ourselves and our world from a danger which one extended aspect of humanity has made clear.

Nuclear weapons, as Lifton puts it, "create a universally shared fate—a technologically imposed unity of all humankind."[14] A threatening technique is not to be overcome with new technical measures. The overall human sense of self is changed as we grow to recognize the vulnerability of our position. For many, the first tenable response is complete despair and the lack of any sense that the world is worth living in, since the danger is so acute and the distance so great between most of us and the decision to deploy these devices. But this despair is only a place to begin. A technology that reveals a human limit in the manipulation of nature demands a unified responsibility from all those who come to know it.

Hans Jonas writes that evil in the world is so much easier to grasp than the elusive good. In the same way, we immediately grasp the menace of total destruction without knowing exactly what will be lost. The contemplation of the suicide of humanity and nature at once requires us to determine exactly what about our present and potential life is worth preserving. "We know the thing at stake only when we know *that* it is at stake."[15] We learn with the Bomb that increased power can lead to greater fragility as much as more strength. If war is made ridiculous as the advance in weaponry hits its limit, then Machiavelli's advice no longer holds: a ruler has a lot more to think about than war.

The intention to destroy itself now needs to be humbled. Nuclear weapons represent the culmination of this ancient human desire. As we have realized this end, we soon grasp the terrible reality of the possession of this power. It is not enough to hold back, to attempt to draw an arbitrary line, saying "we will use weapons only as dangerous as X and go no further." It is more important to reconsider the entire course of reasoning which leads to war in the first place. Concentration on the bleak imagery of nuclear holocaust suggests ultimate answers like irrevocable despair or postmortem resurrection. Specific kinds of technologies might be banned without forcing us to deal with the underlying problem. Many more people died in the gas chambers of the Nazis than at Hiroshima. It is not the technology that is as frightening as the measures of cruelty acceptable by those whose highest intent is eradication of the enemy, without regard for the humanity which all of us share. This total destructive mission can be no more. Human life can no longer be conceived as a state of perpetual war between opposing factions, with rare moments of temporary peace. The kind of international cooperation that has grown astronomically since World War II implies some acceptance of the unity required by our recoil from a perilous technology, and some questioning of the intentions that led to that technology in the first place.

Can the technology itself be redirected toward peaceful ends? Splitting the atom is neither absolutely evil nor the source of unlimited nuclear energy to end all power worries. Historian Spencer Weart asks for "a moratorium on the archaic images that incite such emotions."[16] The mushroom cloud should not be manipulated into a mere propaganda piece for doomsday or the hubris of humankind. It does not necessarily symbolize either the end of the world or human mastery of the forces of nature. Closer to our own acts, the Bomb reminds us of the need to balance our own capacities, such that none of us exists to do what *can* be done, but only what *should* be done. It sounds a call to recognize that technology's real power lies in simultaneously making nature and being guided by nature, letting us live in the world without destroying it.

UNWANTED HEAT

In the early 1950s, people across the globe were convinced that the testing of huge hydrogen bombs in the Pacific ocean was adversely affecting local weather patterns, delaying the arrival of spring and causing repeated storms or drought. The overall public perception was of *unintended* contamination: the bombs were set off by people who did not really understand the ramifications of what they were doing. When tampering with a natural order, adverse effects cannot be predicted.[17] The weather has long been taken for granted as something we may talk about, but not *do* anything about; now its unpredictable independence was widely called into question. As humanity experimented with a new, destructive tool, we endangered what we did not understand.

Unexpected and unpleasant changes in climate contributed to the public outcry to ban nuclear testing. People began to realize that technologies of various kinds could have adverse effects on the natural environment quite distant from the effects they were intended to have or the ends they were meant to realize. This helped to inspire the worldwide environmental movement of the sixties and seventies, in which all kinds of technologies were questioned in the name of preservation of the Earth and return to an Arcadian natural balance among all species which humanity has upset. The horrors of pollution never appear as dramatically as the explosion of an H-bomb, but their prevalence and closeness to the heart of industrial culture make them much more immediate and harder to ignore.

Until the 1980s, the ecology movement lacked any technological ultimatum to unify itself around. But with the recent slew of empirical evidence in support of the greenhouse effect, this may be changing. If the Earth as a whole is gradually getting warmer as a result of our cumulative technological activities, a tangible, if still distant limit comes into view.

The term *greenhouse effect* was coined by Jean Fourier in the nineteenth century to explain how the terrestrial atmosphere allows sunlight to enter while simultaneously absorbing the heat radiated by the Earth's surface. This warming phenomenon is what makes the planet hospitable to life. It is first a necessity, not a danger. Problems arise only if the warming effect is too successful: if increased by either natural cause or human meddling, the world's climate is sent into upheaval.

As early as 1896, chemist Svante Arrhenius considered what would happen if coal plants continued to release carbon dioxide into the air at then-current rates. He hypothesized that the temperature could eventually increase by as much as 9 degrees Fahrenheit, a figure not far from the 3 to 8 degrees suggested by climatologists today.[18] Such a change might radically transform the ecology of much of the globe. According to current predictions, nothing conclusive will be felt for at least fifty years. So it is difficult to provide empirical proof of the accuracy of the hypothesis. Evidence for and against global warming is presently garnered from the study of the few accurate climate records, present only for the last fifty years, with an increasing reliance on computer models that play out various temperature change scenarios and forecast the range of possible effects.

This is not the place to debate the validity of these studies. What is most important here is to note that a majority of scientists have recognized the seriousness of the greenhouse problem, and yet it is nearly impossible to argue for its validity upon traditional scientific grounds. The ultimatum of global warming is something that will only come several generations into the future, and there are far too many unknown variables involved to convincingly predict what will happen with the certainty demanded by science. By the time we know for sure if the predictions were correct, it may already be too late to curtail massive climatic change. To solve this problem, we need to begin to act long before we are certain there is a problem.

It becomes a matter of faith, not just empirical investigation. Belief in the greenhouse effect lends support to the warnings of environmentalists which began thirty years ago. Now we have a clearer idea of what will befall us should we *neglect* to save energy and reduce the indiscriminate waste of resources. Environmental crisis seems more like a fact if a single, simple variable like global average temperature can appear to serve as its measure.

But the consequences for assessment of technology are even more serious. If the "success" of the atomic bomb reveals that our intention to destroy is misguided, then the cumulative human emissions into the atmosphere which augment the greenhouse effect suggest that even technology that does not intend harm may be part of an unplanned global scheme of devastation. Technology is the primary means for making the world into a human place. If this process inadvertently destroys the world, the entire human project is doomed to failure.

No species was meant to live forever. It may be a uniquely human tragedy to be the first species that discovers this about itself. And no species easily thinks beyond its own opportunistic survival. We are the first of whom more than this is required, because collectively we have adversely altered the atmosphere of the Earth. No one technology can be singled out as scapegoat for the rise in temperature, and it cannot be announced to be one individual's fault. The scariest thing about the situation is that the totality of technologies has caused the problem. Every little change contributes. We are all implicated by this push against the planet's limit to accept us. Global warming is a warning of a destruction caused by everyone, intended by none.

We need to *believe* in this problem in order to solve it. Data cannot be collected about the future, and past records are really no help in predicting the outcome of a world so galvanized by incessant change. Critics fault greenhouse effect research for relying too much on computer simulations, which only play out the consequences of axiomatic premises that are always somewhat arbitrary. In other words, the data for the future are the result of an intricate nest of calculations. Any correspondence to the real world will not come to light for fifty years. These models are an elaborate form of hypothesis, built on faith in an Earth deeply tied to all our actions. A sense of ecological interconnection precludes any awareness of crisis.

The distant future lies by definition beyond the reach of empirical observation and the collection of data. Technology causes and suggests the possibility of a problem. Response to the threat needs a motivation apart from technique. Jaspers tried to place humanity midway between necessity and evil, in order to argue for the inevitability of the ultimatum of the invention of the Bomb. With the greenhouse effect, the technology necessary for human proliferation turns out to seem evil with respect to the Earth, as it portends global damage with the accumulation of its use. Need we conclude that this danger is an inevitable outcome of the human project? Or is it alleviated if this project is defined as making a home for ourselves in this Earth, rather than aiming for mastery of our environment?

The moment of questioning is inescapable, but the uncertainty is a phase, not a final resting place. It is important for us to have come up against the Earth's limits, to perceive the precarious fragility of our uniquely human way of adapting to the world by molding it toward our needs. This is the most contextual definition of technology: *the humanizing of nature through the refinement of tools which increase the hospitality of the Earth.* When the totality of these efforts endangers the entire context of human life, we begin to question the whole enterprise. And this period of questioning is a necessary step in learning how to temper our techniques. As the Earth gets hotter, we are consistently reminded how necessary it is to examine the more distant implications of each technology we put into practice. Awareness of our impact requires us to look further into the future, and farther away from our immediate lives towards a global responsibility of which the atmospheric thermometer may be the measure of our neglect.

Lifton considers the Bomb to be the fact that finally unites humanity around a "species self" that implores every one of us to accept the necessity of preserving the one living species that has learned how to completely eradicate its environment. Yet the unintentional danger awakened by the greenhouse measure is in some ways more disturbing. It suggests that it is not enough to withhold our urge to kill each other, since even our peaceful life injures the world, and consequently ourselves. Saving ourselves means saving our world. Today the problems cannot be separated. Any realistic environmentalism will also be a form of humanism. The Earth is now as dependent on us as we are on it.

Bill McKibben dramatically announces that our time brings with it "the end of nature." [19] This seems too strong a formulation, given the tenacity of the world and the limits of our own ability to know very much about it. Only one specific vision of nature is called into question by our discovery of greenhouse troubles: the comparatively recent idea of a nature large enough to envelop us, while remaining distant, mysterious, and fundamentally beyond the content of our gaze. This romantic view is what is presently in trouble. It needs to be replaced by the idea of a nature that includes us, which we understand to the extent that we can find a *home* in the surrounding forces which respond to our collective actions, just partially in our control, however far we try to bend them toward us.

Our conceptions of nature depend upon us long before the greenhouse effect is discovered. The discovery of human influence on the Earth's heating system only reiterates the fact: we will never know any such thing as pure, wild nature, empty of human conception. The moment we identify nature as such, it becomes ours! The minute we call some area of the world separate from our influence, we constrain the environment. We block it out. Attention to wilderness is a consequence of a civilization that sees itself as detached from nature.[20] As cities grow to seem insufferable, we long for a pure idyll outside their walls. But in the greenhouse scenario, this Eden is no more. It was always an illusion, and now it cannot even appear convincing. Everywhere the climate changes as a result of human presence—inadvertently, unplanned. If we choose to hang onto an idea of nature, either as overall context or elusive goal, it must be a different idea than this.

The fortunate side of the greenhouse effect is that it graphically reveals the connection between humanity and all conceptions of nature. Any new view of nature now involves some kind of a return to the older idea that we are as much a part of nature as it is of us. When we act correctly, we want to believe we are following the authority of nature, as Aristotle repeatedly emphasized. When it succeeds, technology reinforces our sense that the world is a home. So it too is mediated by nature, even while continuing to alter the meaning of nature. Nature is no easy or given fact of life for opportunistic, inquisitive humanity. To get there is hard work, like taking the trail to any nameless destination. In Lao Tzu, the Tao is a flowing state which the wind and the trees already know, but which humanity must strive unceasingly to achieve.

Our road toward nature also includes realization of the difficulty of the path. Technology begins to prefigure our lives as soon as it embodies our surroundings. The techniques of city, asphalt, and wires bearing energy and information have become the defining attributes of our environment, a long way from tools that directly extend the hand. Today, evidence suggests that the total technology of the world we have made may be insupportable, and eventually contrary to the interests of all species on Earth, including our own. As with the Bomb, we want to question the legitimacy of the intention that guided the development and proliferation of the human-made problem. But we are unable to find any single intention upon which to pin the blame. With each we only wanted to make the world an easier place in which to live. The inadvertent destruction in which we are now caught up is like an explosion unfolding in slow motion. We all have our fingers on the button, testing the resilience of the world against any plans to improve it.

The entire intention that comprises the movement of humanity is called into question. A fatalist answer might be that we have already had our time on Earth, and there is no reason to imagine our ways are sustainable enough to continue. An opposite response might emphasize that we have always mastered enough of nature before to endure, so why not expect the same in the present quandary? This route forgets that nature is changed as we claim to control it, and that technology rarely offers windows to glimpse possibilities that lie outside its rules.

We need instead to strive to fit in with the world. This cannot involve a return to an idea of nature which once simply included everything we as a species are able to do. We now know too much to believe that. We want to be accepted by the world even after we admit that our way of life could destroy it. Technology has extended the limits of our responsibility because it enables us to make greater mistakes. The more powerful it becomes, the more it demands a response outside the technical: the extension of care outside the realm of our own limited actions. As these actions combine to test the limits of nature, we need to consider ever wider interests of the world as our own.

Lifton wants his "species self" to respond to the atomic bomb, so that the urge to destroy wielded by the hands of a few does not eradicate all of us. With the unintentional devastation in which we all participate, even wider consideration is needed—care for a self that includes the full world of human and nonhuman realization. The more our collective life becomes the extension of ourselves into the world, the less possible it is to imagine any human interest apart from the interests of the Earth itself. This may be an ancient premise, something common to the outlook of many primal cultures, but for the culture that defines itself by successively "improving" its environment, it only begins to appear on the horizon of a time in which the malfunction of our efforts becomes apparent. Through technology we have increased our dependence on the world rather than achieved any imagined independence. We require *more* resources than any previous civilization, so we need to care more about the sources of energy in order to preserve them. At the same time, we need a goal greater than increased consumption to fuel the development of our society.

Extension looks like a convenient metaphor for a dynamic way of life which goes nowhere in particular. We start with the human hand, and propel our forces outward. We expand into the world without needing to choose a direction, simply increasing our dimension as we emanate from the starting point of ourselves. But the grand success of technology in this century clearly shows that this expansion reveals limits inherent in all human enterprise. As we continue to reconceive nature, natural limits crop up to remind us that our intentions are tempered by the world that makes them possible. And as we endanger this nature, our intentions themselves need to be tempered and refined.

So how does nature remain a goal? Technology claims to liberate us from the constrictions of a hostile and fierce environment, while in fact it ties us more tightly to the ecology as we extend ourselves through tools and techniques. To be human means to be extended through technology into the world, defining nature along this course. As we try to bend the world toward human need, we find ourselves closer to those parts of it that seem unwavering "facts" of nature. We need more resources, so we must protect more resources. Only after being propelled far by technology are we able to pause and look back.

Technology appears to increase the dimensions of humanity without end. But today it runs up against a wall of resistance. The greenhouse effect bespeaks a total earthly boundary to the proliferation of technical improvement. Rather than turn back from this edge to the dream of some pastoral return, we need to restructure technology so we can continue with it. It must be conceived and constructed so as to encourage us to adapt to the world it has opened for us, rather than bar us from that world's potential.

THE SAFETY OF DISTANCE

Having learned to make use of the elemental forces of physics to produce our own explosions, we can be said to be *closer* to nature in the sense of controlling the interaction of elements. Yet do we know enough to take responsibility for each catastrophe, to know when a Bomb should be let loose upon the world? A similar kind of closeness encroaches when we each accept some responsibility for the overwarming of the atmosphere. The images of a heated Earth are extreme and frightening: the flooded metropolis, the rising ocean, the melting icecap, the spreading desert, the endless heat wave. And the simple technical measure of temperature shocks us into admission that the danger is real.

Ought we follow Weart and shun these utmost environmental images along with those pictures of nuclear devastation, unhealthy because of the hopeless paralysis they inspire? The extreme also incites *concern.* The mundane, difficult changes in machinery and lifestyle which hold the answer do not. Perhaps the current fascination with ultimate images of elimination is something of an aesthetic experience, a feeling of awe that comes with the confrontation of dramatic and possible ends to our world. Immanuel Kant called such a feeling a judgment, a decision, an evaluation of the infinite in opposition to humanity. In his view, we appreciate the earthquake or the violent tempest only if our own position is secure:

> We readily call these objects *sublime,* because they raise the forces of the soul above the height of vulgar commonplace, and discover within us a power of resistance of quite another kind, which gives us courage to be able *to measure ourselves against the seeming omnipotence of nature.*[21]

For Kant we are moved by such upheavals only because we are detached from them. They are important solely because they remind us of the infinite greatness of our own mind. Pure reason is the domain of the *beautiful,* while nature knows merely the *sublime,* a lesser category by virtue of the fact that it is knowable only in opposition to humanity, never purely in itself.

Today we cannot afford to be so innocent. The stark ultimatum of a nuclear explosion or an Earth in climatic imbalance are deeply tied to the core of humanity, never opposed to it. Kant preferred inner, purely human beauty to the extreme beauty located within nature, because the latter always included some amount of horror in its distance from humanity. But now the most extreme edges of the natural are bound to the acts of humanity. We do not retain the luxury of admiring nature at a distance. The world must now seem beautiful or terrible with us, not against us. We need it more than ever to survive.

The world as a whole must be considered to be a home even after we have shown the potential to destroy it. Perhaps "home" is a more tangible idea than the all-enveloping weighty abstractions of "nature" or "world." Architect Christopher Alexander points out that a true sense of home may be built only once one admits the reality of impermanence and of death. He writes that if we are committed to "creating nature in the world around us, we cannot escape the fact that we are going to die."[22] Although he is most concerned with finding the rules for the construction of specific buildings that flow in a timeless, nameless, natural way (see chapter 6), the idea becomes more challenging when applied to shaping the entire planet into a home. If we form the Earth into a place eternally designed for the promulgation of humanity, we neglect the ebb and flow of life. Does this mean it is better to live with the knowledge that our time on the Earth will be limited, temporary, passing and finite?

Do we prepare for the end, or move to tread lightly? After the approach of intended and unintended desecration of our world, we know the complete human tragedy: to have understood and conceived enough of the Earth to perceive that our way of life endangers the rest. This realization is potently described by the late Norwegian philosopher Peter Wessel Zapffe: "What we call nature shows neither morality nor reason—Its degradation is inevitable, and nothing, not even the most glorious of human achievements, can escape final annihilation."[23] Zapffe believes it is the human gift and curse to imagine morality and order in a senseless cosmos. Our tragic flaw is simply *prowess* in transforming the world. Now we turn to assess what we have accomplished, and conclude that the world would be better off without our kind. Thus we are the only species to recognize our own redundancy.

For Zapffe, there is only one solution: abdication of our reigning role. Let the species die out gracefully by refusing to reproduce. It is a vision hopeful for nature, hopeless for humanity. Few may follow him this far, but certainly the gravity of current problems lends support to his choice. It demonstrates how paralyzed we may feel after considering the whole situation, after taking in the perilous limits that humanity has unveiled in its investigation of nature.

Resignation comes when we allow ourselves to be swallowed up by the immensity of the problem, knowing it is so much easier to destroy than to learn. Though we may all be implicated by the extremities of human error, we need to find a way to focus on the specifics of our own actions, rather than on the stagnancy of universal malaise. Sure, the world as a whole may seem impervious to reason or the true. It is too large for that and we are too small. We locate order in those systems determined by what we have been able to construct—either those things we have made, or those parts of the external world that resemble what we have made. The unknown remains beyond our limited categories, yet we cannot let our discovered limits destroy it.

Humanity as a whole will not likely sit back and accept defeat, even if evidence appears against us. We are active and opportunistic, and will wish for a course to continue on after accepting the immanence of danger, not by rejecting or denying it. Our actions are direct, never as global as the overall peril they have revealed. This is why it is important to develop a conception of technology that examines how the individual technical act operates in response to the threat of the whole. This is why it is necessary to return to specific instances of tools that enhance our position in the world with an enlightened response to destiny.

It is as a totality that technology has come to seem evil, damaging the very humanity it was invented to extend. This chapter has presented two of the most extreme instances of this dark side: first, a destruction planned for strategic purposes, revealed to be more significant as symbolic of an uncontrolled side of human purpose. Second, a warning of a global heat wave, suggesting that the sum total of human improvement is self-interested and blind to the context that permits humanity to flourish. The sheer magnitude of the devastation suggested by each supports a pessimistic outlook. If it is not humanity that has proved itself unworthy, perhaps it is modern technology as a whole. If it has reached limits that call everything into question, perhaps the technical might be isolated as a disease, and eradicated from infected humanity.

This is what thinkers like Jacques Ellul suggest, condemning technology as a force that "resembles only itself, with its form and being identical."[24] Automatic, self-augmenting, universalizing, monistic, and ever-expanding, the technology of our time seems to swallow all processes and activities in the service of an unprecedented growth whose limits prove that *la technique* lacks any morality whatsoever. How else could it engender such ruin? But, as should be clear by now, such a monolithic vision of technology is of little use in any future reform of specific technical cycles. *It* does nothing. We are responsible, and need instead an individual response to each situation, one cognizant of the dangers observed in the whole.

The "positive peace" wished for in opposition to total, active eradication perpetrated by nuclear war is the establishment of nature as our home through the individual exploration of technologies that extend us by placing us further into the world, not above it or in control of it. Whenever technology is working, we should feel it as extension rather than as something that enslaves us or suggests our mastery. It must bring us closer to the Earth, by ensuring our fit. It should open us up to the environment by heightening our dependence on the surroundings. It need not hide the world behind the mask of an artificial order, a human grid as simple or right-angled as we want the world to appear. We do not remake the environment in our image, but extend ourselves to match up with an evolving nature.

The next step is to connect this overall imperative with the cycle of individual technical intention, realization, and renovation which is the central metaphor of this work. Technical changes are instituted through the solution of individual problems with the support of political processes, not through the adoption of enveloping and pervasive guiding philosophies. The task is to temper the effect of a technical success on its initial intent with care for the sense in which each particular invention contributes to the overall limits that have been discovered for technology as it tests the tolerance of the world. Every single tool can be examined not only for ways in which it extends humanity, but also for what it teaches us about the demands we make on nature.

For as much as techniques reshape the way we conceive nature and the universe as idea and system alike, they also reveal how little we are without the world around us, as context and limiting force. When technology threatens to demolish this world, either on purpose or by accident, it has decidedly failed. We will need to consider whether each innovation cloaks the context with its solution, or unveils unprecedented ways in which humanity and nature are enhanced by the same new opportunities.

Look at the bridge scissoring toward the decapitated walls of the enemy. Think of the destruction of home, of the invasion behind shattered walls. Our time is like that prophesied at the conclusion of the *I Ching,* that ancient Chinese answer to Spinoza's geometric system of human behavior. Here is a taxonomy of human situations contained wholly within arrangements of six lines, called hexagrams. The final hexagram is entitled *Wei Chi* (☲☵), "Before the End." This is the necessary close to any query of human fate, with us poised at the brink, ready for the worst, prepared to change it:

> The time *Before the End* can be compared to a lengthy trek over a high mountain. At some point, before reaching the peak, you can see in detail exactly how much further you must travel. You will know what is involved in reaching the top because of your experience in the climb thus far. However, when you do reach the peak, which has been in your sight for many long days of effort, you will have done only that. You will have no experience whatsoever about descending the other side. To rush up

and over the top in an overly confident manner could bring disaster.[25]

Now there remains a tension between danger and peace, between discovery and home. Go one step *back* in the *I Ching,* and you will find hexagram #63, *Chi Chi* (☲☵), "After the End," representing terminal balance, tranquility and settlement in a state of perfect equilibrium. This is not the final resting place, but the next to last step on the way of human change. Why can't we end with serene fulfillment, at one with nature? The end is never the final chapter. *Preparing* for the end is.

A strange image of a home, with water as strange prisoner. Housing nothing but a machine, it is a mill that blocks a stream, channeling the current into energy to grind. Home on the river, movement flowing through it; trees and plants around. The walls and roof contain our use, yet birds swim in the measured pond. A diversion of nature—does it control without deceit?

What is the power of the warm enclosure made by the walls of home? As the place where we and our machines belong, the bricks open and merge with the garden as the meaning of Earth becomes their limits.

Technology changes the meaning of nature as it continues to seek nature. The virtualization of technique which begins with the wheel and ends with the computer successively brings the notion of tool away from direct contact with earth and towards complete imitation of the universe with the set laws of logic. And the culmination of physical technical prowess is the warning of immanent destruction of the planet, either intended or accidental. These inventions move us away from the ground, but we should return, asking the following question: "What kind of technologies instill us in the world?"

Not those which give us control of nature, nor those which remake nature into our own image. We search for those tools which bring our own interests in line with those of the larger world. As we learn more of nature and extend our fit therein, the limit of humanity is extended beyond the narrow concerns of any individual human life. The goals of the world are also ours, but only if we can find definite tools and specific instances of their use that suggest that extension takes us somewhere.

THE TUNNEL UNDER THE WATERFALL

If nature is a goal so essential but also elusive, can it offer any guidance? I offer here an account of a trip from perhaps the most plastic to the most profound sides of the machine.

Through the maze of what must be the world's largest gift shop, you will find the entrance to the tunnels under Niagara Falls. It is best to visit on a rough, stormy day, when the entire neck of the river is shrouded in mist, so it is impossible to see anything above or below the waterline. On a day like this, the management will most likely discourage you from descending the elevator into the tunnels, claiming "there is nothing to see," but if you insist, they will have to let you go. Sealed inside complementary plastic raingear, you may be lucky enough to descend alone, so that the full magnitude of the technology will be apparent. Walk straight to the end of the watermarked stone corridor, turn left and proceed to the edge of the abyss, where only a slight railing separates you from the onslaught of rushing water. It does not look like water, or even feel like water. Most present is the tremendous rumble, a roar of the flowing Earth. This precedes all other perception. Gazing into the whitewater left by the sound, you see abstract patterns in foam and froth. White tongues, white swirls, some of which splash into your face to remind you that it is a waterfall in which you reside, no pure design of noise and light.

It is nature, not art, which is apprehended through the white fury at the end of the soaked shaft. But it is only the artificial pathway, cut and blast into the core of the rock, which makes this experience possible. A violent, rough course of dynamite and explosions made this route into the heart of Niagara. Once inside, we are able to perceive an unceasing force beyond our power to create, an endless surge which is not judged with the safe distance by which we may choose to view a work of art. It does not seem to matter that the entire flow of the water is regulated and controlled. This is an unstoppable force, always thundering and shaking with or without a person to notice it. The technique of the tunnel has brought us closer to it. If we go down there, and are able to shut out all the tackiness and kitsch which has been built around the "attraction" of the Falls, then we are genuinely closer to nature. This is a technology that works. Whenever or wherever I now see water, I conceive it differently after I have sensed the possible magnitude of its power. We do not grasp the significance of the Falls until we have extended our reach towards them, so that the water in its might may touch us and envelop us.

Yet it still does not destroy us. It does not even frighten us, as we stand in awe of it from a vantage of relative safety. This is the point Kant chose to emphasize in his *Critique of Judgment,* as he attempted to analyze such aesthetic response in the wake of natural magnificence, writing at the dawn of the romantic age, when such spectacles as Niagara Falls began to be generally accessible. The last chapter mentioned why such natural wonders were deemed sublime because of the great gulf that stands between them and the crystalline kind of beauty possible only within the human mind. The sublimity of the violent workings of nature is only apparent in contrast to human interest. These powers would in direct contact demolish us, but we enjoy them if removed to a place of safety. It is "nature beyond our reach,"[1] enjoyed as if an ultimately powerful artistic statement.

This is what Kant thought at a time when the enjoyment of nature began to compete with the enjoyment of works made by humanity. Today's situation casts the problem in a different light. We have by now built a whole culture out of the turn towards the romantic, which seems also to deny the purity of any aesthetic assent to nature with a different argument: there is today no part of the world independent of human manipulation or selective gaze. We see what we want to see, we make what we think we need.

The tunnel beneath the Falls suggests that our acts are neither opposed to nor determinate of nature's. Closer to the total limit, we treat the part with greater respect. This underground causeway is no great upset to our environment, and it does little to overtly reinforce the human stamp on things. Yet it brings infinity closer on all sides with the touch of thunder and mist. As we step to the end of the subterranean corridor, we are released, through technology, to the world.

RELEASED TO THE EARTH

Leading us inside the roar of nature, the tunnel under Niagara is made part of us through our own construction. It is a poignant image for what Heidegger calls *releasement,* a concept which is his most evocative final comment on how technology should be improved in our time.

The German word is *Gelassenheit,* whose usual modern meaning is a kind of composed and calm indifference to one's immediate situation. But Heidegger looks back to its etymology, and reconstructs an active yet accepting process that allows an individual tool to let its user flow into the world, becoming a part of nature through active engagement, *not* severed contemplation of the awesome and frightful.[2] It is a slogan to help find those techniques which connect us to the larger world by letting us succumb to the powers beyond our capacity to calculate, order, or judge. The world we are introduced to is multiple, as the successful technologies offer "releasement *toward things.*"[3]

This releasement is Heidegger's way of accepting and rejecting technology at the same time. He does not give in to the technical, nor condemn it as human intrusion upon nature. Technique is neither the answer nor the problem, but a fact of human life. If we assess it and treat it correctly, it may show us the way to realize the human goal of fitting into the world while letting the world reveal itself through us. This is his cryptic response to the situation created by Enframing discussed in chapter 3. Technology need not preface all, but it should leave gaps where essence may shine through.

The idea of releasement was conceived around the time when many imagined atomic energy would offer, in the near future, a wholly unproblematic kind of power source that is not extracted from the Earth like all previous sources, but concocted by human ingenuity out of the fusion of minute particles of matter. It seemed to promise, at the time, unlimited freedom to expand electricity, without offering any advice on what to do with this newfound independence. The planners for the future simply assumed that we would need as much energy as possible, so they pursued the nuclear alternative with vengeance.

We would become caught up in the new, addicted to the inevitability of mechanical improvement. It is easy enough in this way to be guided by the streamlining of technique, as there is always a smoother and stronger new approach to be tried. If energy actually became as plentiful and painless as scientists dreamed in the 1950s, would we simply take it for granted and lose any respect for the cost-cutting and the efficient? Probably. And the sheer opportunity would make it even more difficult to ask the question "why?" in response to a suddenly possible machine. If it *might* be there, we will pursue it. Possibility is enough.

This is the danger of a purely technological rationale. It moves us forward, but does not bring us closer to any goal in itself. Only from outside of technology do we discover how the tool contributes to overall human purpose. The technique is not, however, a simple means to an end, a solution to a problem. It enamors us of the world, sets us into our place, and points us toward further achievements. If it works, it creates a home for us. If it is overdetermined in a single dimension, it cuts a swath through all alternate possibility, and sends our vision down a tunnel, one with no gushing spray or surprise at the end, but only light at our feet, marking the path, silencing any question of it.

Releasement as a goal is equivocal. It suggests that it is good to find tools that open us up to the world so that we sink into it, without falling passively into an empty canyon. Heidegger is asking us to let go, yet to select the course we are given as if blown to a destination with a billowing sail. This is evocative, not definite. He is trying to describe the world as a place that offers us direction *without* resort to the troublesome concept of "nature" which has been shown to depend so much on the techniques we use to approach and delimit it. In his conception, technology does extend humanity, such that we come closer to *something*. The goal is not to be falsely pinpointed with an ambiguous embracing name like world, context, or universe. It is only the *distance*. We are released to the far-away.

We are meant to live in a way that brings us closer. This life's essence is affirmed by a renovation of the concept of thinking itself, recalled by Heidegger as "coming-into-the-nearness of distance."[4] This thinking is both place and activity, noun and verb. Thought is action, because it touches the distant world with tools, while reflecting on these tools to consider where it is that they bring us. When we think we wait. We wait for sufficient openness to the surroundings. As we release ourselves to the enveloping world, we change it by letting it appear to us.

Skeptics may accuse Heidegger of merely "moving freely in the realm of words."[5] But he would say this is the strength of language, precisely what it most allows us to do: to play, to test, to reach towards the change in our context before we actually build it. Conversation (and he would never say argument) sets up the instructions, dancing around the action of releasement. We need to imagine the goal, yet we cannot express the goal. It is all around us, and still too far to be seen.

Technology may help us approach this goal of the world, or it may help us forget the goal, as it brings its own insurmountable ethos of technical improvement, an unending search for efficiency. This is why we must be careful with the new tools we invent. Each of them contains enough to lead us astray, and offers nothing to hold us to our path. That must come wholly from a self-reflective humanity, and we are increasingly defined by the world which tools disclose.

Heidegger tries a succession of images to instill in us a permanent fix on the distant reality. He ends with the shortest, most apparently incomplete fragment of Heraclitus, number 122: *Anchibasíe*. Only one word, usually translated as "dispute."[6] For Heidegger the moment of disagreement is most significant in the fact that in the midst of it, one turns toward the opponent, facing the adversary. The word is then rendered as "going toward," or "going near." We need *to weigh the pull of opposites* before we approach anything resembling our rightful place. We take on the Earth, and do not simply settle into a comfortable seat. Letting ourselves go to the precision of proximity—to what is ours and the world's and never simply "nature's." Heidegger is wholly concerned with defining humanity, a task that only comes to fruition with correct care in our inevitable manipulation of the environment. Do not think of our work as renovation or inhabitation, but instead, as capitulation. We work hard to come into our own. This way the elusive coalesces right before our eyes.

Releasement is a message enshrouded in mystery. Heidegger no doubt wanted it that way, so that the usual, intrusive way we speak of technology as a vehicle for mastery of the Earth would be challenged. The search interprets the world from the ego looking out: I choose the tunnel to the cascades; it does not make the water mine, but brings me inside the water. As long as I remember the time there, I belong to it. This I owe to the passageway, a technique intruding deep into the ground, but bringing me close to an awesome distance, a destructive force which I allow to touch me. It does not recede even after I have left it, and I hear its roar even now. The tool has changed what water means.

At a loss for words, with the demand to name, I am tempted to call the source of this force "nature." But nature remains more than the sublimity of thunderous water which technology can help us approach. If it still is to be the overall context for human activity, we need to consider much more complex relations between the human and the natural, which demonstrate how *both* are transformed as a result. We are released toward an end that alters as we near it; surely this is what Heidegger realizes as he *avoids* the halting term *nature*.

Technologies make specific parts of the world accessible. They focus our view into limited places. Can they then be said to open us to a single, whole "distance"? It seems that they select certain horizons for us, thereby making others invisible. So there needs to be a way to translate abstract releasement into a strategy for the concrete.

THE PULL OF OPPOSITES

The tunnel takes us down "into" the falls, but with its success, the water's heart could be said to have been blasted out. Nature itself is changed and remade there, as we are permitted to climb deep within it. Will we still fear the torrent in the same way if we may benignly touch it? This depends, once more, on how absolute we consider the goal of our approach to be. Is it free and wild nature we have dug ourselves within, or has the wildness been tamed in the wake of invasion?

One way to recommend a technique is to wish the tool to be *transparent*: once learned, it should directly take us exactly where we want to go. We affirm the desire, "pick up" the tool, and without afterthought solve the problem. There is no need to think of the technique as long as we are engaged with it. This kind of tool is invisible, as we realize a pure extension of our mind/body in the process of intending to act—as we forget the tunnel except as conduit to the light of white water at its end.

On the other hand, the more powerful a technique, the more it re-makes the world in the pattern of our wants. Chapter 2 demonstrated how the more enveloping techniques function less as a direct means to realize our desires, and more as immediate reconstructions of the world in the image of human order. These tools not only remap, but rebuild the world, prefiguring our physical and conceptual environment with the rules we would wish it to have. These tools are *transformative,* changing the ground as we figures are propelled within it. Our distant context, now brought nearer, takes on the qualities of regularity and efficiency which are seen as virtues from the mechanical vantage point. They are loved because they can be dealt with—part of the solution, not the mystery.

Both these notions, transparency and transformation, clearly seem to refer to someplace outside the technology in which to locate its value. They open up to a whole series of dichotomies that try to divide technology into opposing camps. What does the machine change beyond the extent of itself? The human intention that drives it. Does the evaluation of technology then become simply an assessment of purely human motives, seeking out the ethical direction for people's lives? This depends on whether we are able to identify a properly human way of being *independent* of the technical extension that instills humanity into the environment. For if not, humanity in the world *is* technology, with no separation possible at all.

Patrick Heelan puts forth the view that there is an unaided, rightfully human way of being which precedes technical renovation—a way of releasing ourselves to the full 360-degree expanse of the Earthly horizon. Today we are nearly always oblivious to this form of situation, because we are immersed in an order of our own creation which we take as the norm. This is the gaze represented by (and constructed upon) Euclidean geometry, which Heelan calls the "carpentered environment." We live in a world built out of rectangles, with a consistent regularity everywhere, because we have put it there. Our houses are rectilinear, and our continued conception of the Earth as a flat square of a map leads us to chart new land into manageable regions bound by right angles, set up to be inhabited scientifically and statistically. Fields are rectangular, and the intersections of roads most often cross at 90 degrees, with travel mediated by the four-sided traffic light, signaling stop, caution, or go. It is all planned, all built, and it is more or less the same wherever Euclid's influence has had its way.

Such geometry does not reflect the plan of nature, but is an artifact of human culture. It is a constructed part of science, a frame of reference which we have literally drawn, out of raw materials, to hold the effervescent, flowing, uncertain world to a human grid. Now do not think that *this* is the purely human intention, preceding technical engagement, imprinted unequivocally in the human mind *à la* Kant. No—Heelan is more in agreement with Husserl, accepting Euclidean geometry as a "form of technical and scientific praxis, not of visual phenomena as such."[7] Euclid is only to be believed if we possess *appropriately readable* technology: squares drawn in sand, wheels that roll down inclined planes, sundials, planed tables and four-sided skyscrapers. Uniform geometry has been concocted to read the chart of our cut and measured globe. It is the result of a successful and transformative succession of techniques; a theory which is the consequence of one very prevalent way of organizing and setting up the world.

The structures that have inspired Euclidean geometry are so omnipresent that it is difficult to question the validity of the change they have inflicted on our consciousness. Yet if we are able to forget (or transcend) this, a purer engagement with the world might emerge. It would be a sense of fit where nature and humanity seem to blur, though with a world still determined from the standpoint of a panning, searching human gaze:

> *In untouched nature,* among mountains and wild places, in sea and sky, beyond the domains populated with visually accessible

Euclidean standards, the geometrical structure of visual space
may become indeterminate, or more likely tend toward the
hyperbolic.[8]

The allusion is to hyperbolic geometry, a system not opposed to organization, but challenging to the uniformity of even measurement and the
parallel postulate. What is near to the observer may be regularly commensurable, but what recedes into the distance is blurry, less defined,
tending toward the asymptotic limit of the infinite. Such a world would
be different for each observing person. It would be locally structured,
while accepting the inaccessibility of a horizon that concentrically surrounds the reach of human action—always to be taken into account,
never wholly understood or reducible to rectangularity.

For Heelan, this originary world must be devoid of "readable" technologies, or techniques that turn the world into a text only legible
through codes set in place by machines or mechanical ideas. No blue-
prints, diagrams, or pattern-books to be examined and trusted. Here
we would have only foaming, fluxing experience, fading in and out of
focus as we moved. This is his way of humanity unaided in nature.

Can a human world really be imagined to exclude technology of any
but the most primitive, asystematic kind? If it is the person-centered aesthetic that is the authentic human way-in-the-world, there need to be
technologies that can *help* us get there, and not just subvert something
imagined to be purely given. David Abram has written that he can drive
his car to an open field, get out, shut the door, walk out into the open
clearing, and take in the wide, ambiguous expanse of the horizon so
that time and space become one:

> I find myself standing in the midst of an eternity, a time without
> beginning or end. The whole world rests within itself; the trees
> at the field's edge, the hum of crickets in the grass, cirrocumulus
> clouds rippling like waves across the sky from horizon to hori-
> zon. In the distance I notice the curving dirt road and my rusty
> car parked at its edge—these too seem to have their place in this
> open moment of vision, *this eternal present.* . . . Things are dif-
> ferent in this world without "the past" and "the future," my
> body quivering in this space like an animal. I know well that, in
> some time out of this Time, I must return to my house and my
> books. But here, too, is *home.*[9]

Another attempt to surround the limit of distance. What is the advantage of this? Somehow we "sink" into nature without those two most
basic of human organizing forms—space and time. They meld together,

far away at the outer edges of our ability to conceptualize. Far away they are one, yet far away is the only place we find them—and together. Close to us we have concrete actions and measured plans. In the distance we have these parallel, knotted abstractions.

Phrased as something asymptotic, fuzzy, and beyond our reach, Abram's experiment and Heelan's hyperbolic model both suggest that our practical life with tools concentrates on the orderly and the mundane, while the contemplation of the elusive asymptotes of our curious culture requires an about-face from the normal turn of life, not a smooth extension. Why? Because there is no set link between our most prevalent technologies and the circling gaze that gets us back beyond the penchant to divide and conquer. Heelan bolts out of the carpentered world of house, backyard, and city and heads for the wilderness to go back to a primal reunion of the natural with the human. Abram drives his car there, but then turns it off to leave the map behind, returning to the territory. If the way to be with nature is so diametrically opposed to the techniques we depend on to get to the place of pure contemplation, how will it be more than a tempting foil to our ordinary life? For these hyperbolic, person-centered visions to offer any real guidance, they must show a link to technology, not a divorce from it.

Now I write these lines in a mountain hut far above the tree line, gazing out at the rolling tundra beneath the hill on which the cabin rests. Unlike most such houses, this one has two large square windows that frame a stupendous view down over the snowfields and jewel-green lakes. The frame grounds a picture always changing, always the product of a charged link between subject and horizon, figure and ground. The rectangle around me is a shelter, the square porthole a window into a fragment of a world viewed, whose attributes tend toward the hyperbolic aesthetic—the near is clearer, less uncertain. What changes thereof is clearly marked, with more information and less doubt. In the distance, what curves are seen blow consistently with the melding of land and cloud. At the farthest reaches of sight, these two blur with equal impermanence and equal gravity. The horizon is apprehension, art, enjoyed. *Techne* exists even at the edge of my field of view.

The point here is that the rectangle is one of those inventions that now embodies our world. If it is but a convention, it is one of those human axioms which has irreversibly framed our surroundings. And some frame is a human necessity, even as the asymptotes guide the hyperbola in the wilderness, as the horizon remains a circle from the subject at the mountain summit. Even words need some frame upon which to be read—a story has beginning and end, climax and respite. Measurement need not be opposed to the qualities it opens up for us. Quantity needs to leave room for quality, else it will frame only itself. The right angle is useful until it blots out all save its own recurrence. If the view out my window is only columns and squares, walls and corners, then I know a great part of reality has been concealed.

Measurement should not cloak revelation. Is it right to call one transformative, the other transparent? A rectified environment is certainly a changed one, but is the person-centered, distant-fading vision a pure and innocent reach into the unaltered beyond? It too must be a human choice in mode of perception. If we decide it is more "natural," then this kind of nature is simply another human option. Heidegger's releasement remains a human option, not an external end. Perhaps we need only be cautious enough so that further advanced technologies are transparent to the basic dichotomies these two tendencies suggest, such as analog vs. digital, continuous vs. discrete, polar vs. Cartesian coordinates, subjective vs. uniform.

This seems to be the view of artist Paul Klee, as documented in *The Thinking Eye,* his collection of lecture notes and pedagogical sketches from his years teaching at the Bauhaus. His task is to instruct art students in the range of rhythms one can use in one's work, taken from the vast extent of possible visual forms. He too recognizes a real difference between fluent change and discrete change, between continuous, liquid variation of light and dark and the step-by-step calculated gradation between black and white, measured and controlled, with each step recognizable as a distinct zone along the marked path. Both of these are available to art or *techne,* and the creative person should learn to master each.

The first he calls *natural order*: "a living balance between the two poles [of darkness and light]—this is the task we cannot avoid."[10] The idea comes from the predestined human context of the rhythmic ebb of day and night. We observe their continuous interchange, and then strive to gain control of smooth variation in our own constructed things. We feel it on the potter's wheel, in the weight of a paintbrush from light to heavy touch, in the ways a knife can sharpen the edge of a stick. Such tools offer uninterrupted variation, without the need to quantify each increase or decrease in force. The continuous may be the first change to be observed, remaining the hardest to describe. Yet we partake in it before explanation.

Also available to us is what Klee calls *artificial order*: the step-by-step motion between the poles of white and black. The smooth endlessness of variation is gone, and the abundance of discrete possibilities takes its place, "impoverished, but clearer and more comprehensible." Following an intention to explain, we eke out a world of specifics. Marvel at each step along the way, and try to give it a name: "Below, dark subterranean rumbling, in between, the half shade of underwater, and above, the hiss of brightest brightness."[11] To measure with any scale, first make the subject *susceptible* to measurement. And this means identifying a piece of it and shaving this off from the rest as some unit with constant, even character. The "natural" order identifies only through change, while the way of "artifice" sees only constant, even limits; together they mark the continu*al* as opposed to the continu*ous*.

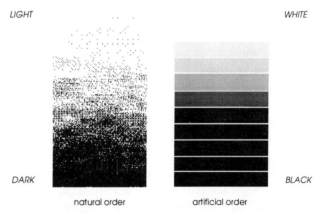

LIGHT   WHITE

DARK   BLACK

natural order   artificial order

Figure 8. Paul Klee's continuous and discrete changes from the dark to the light

Klee juxtaposes these options not to pronounce one superior or more authentic than the other. No, he wants to show that each exists as a means by which we may represent and take stock of the world. Each is a principle with a place in art or design; both may translate or remake the universe. There is no need to call one nature and the other humanity. We perceive both ways, we make use of two tools of thought. *Either* can be near or far in relation to our gaze.

This is a modest way of dealing with the categories "natural" and "artificial"—placing them both within the context of human knowing so that any external mystery may still remain beyond them. Each is a kind of human measurement imposed upon the always unknown context. It might be more instructive, however, to use two terms familiar from the discourse surrounding computers: the *analog* takes stock of the continuous, while the *digital* divides the whole into identical, exact, and even units. Both are part of technology as elucidated herein.

With this distinction in mind, it is possible to consider the six categories of technology in terms of how they accommodate both poles. If the analog tends toward the transparent, it will be a transparency to human intention, not to any pure and preexistent lifeworld known deep inside to us all. If the digital tends to the transformative, it is a transformation of humanity within the world in the thick of its own process. Here, then, are the six categories, considered in terms of analogical flow or digital units.

1. *Strengthening the grasp.* As long as tools of the hand exaggerate or further direct human movements, they guide a continuous intention without needing to break it down into parts. We become the needle as we thread it through the cloth, and there is no need to calculate each puncture of the raw material. However, as soon as we want to make something out of this extending action, we tend to impose regularity upon the task, trying to place each stitch an equal distance from the last. Done by hand, exact correspondence with this digital ideal may be futile, but we approximate the most manageable sense of consistency. Whether the imperfection of the result matters depends on what regard one has for the qualities of the handmade: unique, individual, only perfect in respect for the situation at hand, not to any exact abstraction.

2. *What we set in motion.* Demand for exactitude and ease of assured control leads to this second category, machines which we do not power alone, but still drive. Among vehicles in the largest sense are included all wheels, all machines that need to be steered. The sewing machine ensures a regularity among stitches, even as real world applications might demand intuitive flexibility. But we try to find a regular stitch for each occasion, working in the throes of a digital aesthetic. We trust it. Here, digital machines are those whose wheels *require* tracks, or leave them like the measured stitch. A potter's wheel, on the other hand, dreams of the circle but always reveals a slight imperfection. It demands no specific scale save the proper application of continuous human force. It remains in the analog realm.

3. *A Separate reflection.* Analog to what? Where is the analogy? The allusion has always been to smooth, flowing natural forces, as some ground of variation that we test and temper. When technologies come to embody our world, they are analogous if they fit into what they use—those that gently sample energy rather than wholly extract it. Irriga-

tion ditches, meandering paths through natural passes. When we re-make the world in a planned image, an abstract regularity takes over. The city grid, precise and unbending. The draining of swamps. The transmutation of oil into electricity, a force let loose across our own wires of communication. Though the variation in strength may be continuous, the decisions that perpetuate the chosen pathways are digital, discrete, and detached from our native uncertainty.

4. *Improving the senses.* In the extension of perception, transparency is at first important. We do not want to know the hearing aid or glasses are there as we process what they give us. The telescope and microscope begin to make us conscious of their frames. And the farther we extend our reach, toward the immense and the minute, the more the result seems dependent on the device, so unattainable with our unaided bodies. To the extent that the instrument itself exerts a firm, definitive, unforgettable presence, the experience is "artificial."

5. *Tools of abstraction.* When it comes to guiding the abstract, we find little outside the pure human tendency to structure. How could we even conceive of nature without/beyond us in the abstract? All we know is *our* abstraction, of discrete units, with rules of organization. Only language is nearly natural among our abstract techniques, because its logic lies a bit beyond explanation, yet seems right and consistent as we describe the world with it. It has evolved at a rate closer to those processes beyond our control to which we still need to pay attention whenever we act. It alters continuously, and eludes particularization. The rest of our abstractions are more clearly designs, human impositions, facts which only we would want to know.

6. *Material memory.* These are pieces of experience fixed wholly outside us, and if they seem devoid of abstraction, they retain some fragment of the unknowable. The frozen image seems at first a true depiction, a snapshot of flowing time. But wait—it is the mark of a moment, an instant that as such is purely a human idea. The camera went on and off, a digital act brought us the photography. Even moving pictures are a series of stills, mirroring motion only out of the many and the separate. We organize things to reflect reality, and have found that the precision of the many may be better combined than the effervescent difficulty of groping directly for the flow. We can never step into the same river twice because we *wish* it were consistent. So instead, we re-make the river as a composite with tiny bits of information about it: its temperature, its velocity, width, color, and chemical composition. Of course it will not be the same again.

The computer counts even the tiniest bits of information in sequences or parallel nets to reconstruct the fluidity of the whole. Here at last is where the artificial begins to compete with the natural, where the success of the digital challenges the endless variation of the analog. For when the *scale* of measurement is so much smaller than the limit of discernment, we imagine that the continuous may be represented by a series of tiny separate shifts too small for us to know.

This is where the machine becomes the greatest affront to nature, nearly *erasing* its ground with the spread of its re-creative process. The world is remade in an image which it can consider; because its digital steps are so much smaller than those we can perceive, we imagine all of the world may be understood in this manner. Yet the unquantifiable is not included, only held at bay.

With this end the "artificial" seems to have subsumed the natural, while the analog is but a remnant of unaided, unanalyzed perception and work. Thus digital computers are so much more efficient than analog ones, though they may mirror the world's flow only by incessant repetition of simple parts. Have these categories then been *reversed* because we have so propelled ourselves into the world that we identify the universe only with what we know, not with what contains us? Does nature then become only the quantifiable, with the rest as human folly?

The whole approach of placing technologies within categories is an example of the meticulous, ordering side of humanity, even if the distinctions as introduced here seem so fluid and uneven. The desire is still to place tools upon a map, to imagine a pattern behind their diversity, even if articulated after the fact. In contrast to this is the vision of overall technology approaching a limit, a hyperbolic view with humanity at the center, our immediate realm regular and commensurable, with the distant hazy and uncertain, *until* we begin to challenge it—first with a digital technology that wants to mirror all, and then with physical changes that alter the far away enveloping confines that make the world hospitable. Test their resilience: as the edges of experience are transformed, they appear transparent. See through them to understand how we are constantly engaged in an effect on the world, which doubles back to redefine us.

SLICING THE STRAWBERRY

So even *approaches* to technology may be analogous or discrete, categorical or limiting, immediate or extreme. The chilling cases of worldwide unintentional global warming or escalated nuclear destruction probably do not make good examples of how to temper technique, because they are so unusual. Yet they represent the final reach of the technical, framing the final conclusion implied by *techne* when extended to historical ends. They do encourage a kind of pessimism, which leads to some sense of futility. But luckily, technology means more than its total effect. There can be no solutions that do *not* oppose the moment to the flow, pitting instant against roving consequence.

In his theoretical work, architect Christopher Alexander attempts to get a hold of a nature that comes clear as a goal, but not a distant, romantic extreme. As a designer, he would like us to build according to nature, while recognizing that a normative nature is impossible to describe. What is right "by nature" is for him a quality, something always *there* outside of humanity, and only occasionally present in our particular works or ways of living. It is *the quality without a name,* an ineffable *right* way to connect people to their environment—a new answer to Heidegger's "poetic releasement," Spinoza's "third kind of knowledge," and Marx's "human manner of production."

According to Alexander, the goal of the human builder (of houses, towns, poems, or sculptures) is to create structures that possess this nameless quality, thus sharing with nature a kind of geometry that celebrates pattern while being flexible enough to adjust to each surrounding, each separate moment and place. The human work should not be an independent self-centered structure, but something that displays the long and hard search for an elusive attribute that cannot be described.

Alexander will not cheapen this motive by naming it, but he will *hint* at it by noting that, although nature may be bent towards the structures we extract from it, in itself it never reflects complete precision. For Kant this made human reason more pure than sense, but for Alexander it means that *homo faber* must be cautious about imposing order on the outside world. "Nature," he says, "is never modular. The same broad features keep recurring over and over again, [but] in their detailed appearance these features are never twice the same." [12] The natural offers a *tendency* toward the kind of order that humans love so well, which our senses simplify to register. Or a more favorable review: the diversity of nature exceeds any attempt to reduce it to patterns. It tempts us with method, but never enough to diminish madness. Every natural result is a unique solution to some given set of guiding forces, while the mistake of uniformity is never made. Boredom is impossible outside human works.

We don't have to oversimplify. That is an error, only a tactic to make sense of complexity. Nature may show repetition and outline, but each event remains essentially unique. Waves crash the beaches in rhythm, but not in exact rhythm. We sense their recurrence, and even enjoy that beat of the water upon sand, never just the same. This perfect combination of order and immeasurable diversity marks the natural, nameless quality—a combination that is so simple, yet so difficult for us embedded in our systems and plans to reach. Nevertheless, it is *possible*. And it should be our goal, if we wish to build ourselves with our tools "into a piece of nature."[13]

Once more an echo of an idea that has become a major theme here. Humanity may become a part of this larger world called nature, but it will be an *achievement,* the result of hard work spent suppressing our belief that simple regularity always holds. Our dream of order will always confront an elemental uncertainty, which we cannot see while cloaked inside our best-laid plans.

There can be building in accord with the kind of releasement I felt at Niagara Falls. It will construct the complex while letting ourselves *go.* It will ask for enough belief in an external nature to have us enjoy being bound by gravity, birth, life, death—changes that always elude measure. A certainty of forces determines us, but exactness remains a solely human vice. We will not build monuments to the reverie of order when we accept its range to be inside us, its role to suggest, not to determine. Here I admit the cut and set frame, but know that it lets me see out to consider how the immediate is definite, the distant sublime in its ambiguity. When I look through the glass and see only other windows, mirrors of my stance, then I know belief in order has gone too far.

Alexander's final, most memorable example of the timeless way is not a building, nor any picturesque town or "organic" village. It is instead the report of a memory: a woman slicing a piece of fruit, dividing a whole into distinct but unique parts. The reason for the cutting is *not* to make sense of or quantify the object, but something else:

> We were having strawberries for tea, and I noticed that she sliced the strawberries very very fine, almost like paper. Of course, it took longer than usual, and I asked her why she did it. When you eat a strawberry, she said, the taste of it comes from the open surfaces you touch. The more surfaces there are, the more it tastes. The finer I slice the strawberry, the more surfaces there are. . . . To live like that, it is the easiest thing in the world; but for a man whose head is full of images, it is the hardest.[14]

Perhaps we slice up the world so that we will *taste* more of it, not because it makes any more sense. In thin slices, there is more to enjoy than the whole at once. If no cut is like any other, then each piece of the fruit will be a new taste. Every morsel need not be a binary yes or no, but a fragment of continuous flavor. If the world is divided in this way, we may learn to enjoy each piece of it to the fullest. If this is the manner we work with nature, we will never let our prowess defeat our place.

Here may lie a solution: use the human ability to divide, but not so far as to conquer. Each piece will be nothing homogeneous, but only a tiny fragment of the flow, so that the continuous may be more readily appreciated. We set up limiting devices so that the world comes clearer into focus. As we know we transform it, we do not want to lose what it first is able to offer us.

BLINDNESS AND INSIGHT

Our heads *are* filled with images, full of actions to complete, designs of machines to help us. And images may convince more strongly than arguments. The extremes of these may have touched against nature, but most of the time we strive gently, without overall plan, toward general improvement through techniques. We cannot always consider how each action affects our overall placement of home. There is never enough time, always too much to do. The question quickly becomes this: how to improve specific technologies so that the larger purpose of making our "carpentered" worlds part of nature will shine through? Alexander is poetic and inspiring, but his dreamlike quality is often uninstructive when we get specific. Can the nameless, after all, serve as a concrete guide?

Two approaches, one of Donald Norman and the other of Fernando Flores and Terry Winograd, show different ways specific issues in technical design can be infused with larger aims. Norman, a psychologist in the field of human factors engineering, finds the evocative and embracing style of Alexander enthralling, but "difficult to put into practice."[15] He tries instead to map out the decision strategies that make a specific tool successful as experienced by the user. Winograd and Flores, computer scientists and business consultants, appear to advocate the direct application of broad Heideggerian concepts to the solution of tangible problems in organization and task management, particularly with the aid of computers enlisted in the service of our actions, not as

models of our thought. They retain the mystical whole while representing solutions, and Norman sticks to the mechanics of individual design
problems and the gap between plan and use. In both cases, practice is
the arbiter of speculation.

The two agree that human action, rather than the generic structure
of human thought, should be considered primary in the understanding
of technique. For Norman, a technical design begins with the perception
of a state of the world, followed by the comparison of the perception
with what we would *like* to happen. Desire forms the basis for technical
goals, at the level of evaluation. Then comes the phase of execution:
intention to act, followed by the plan of technical action and the execution of those actions, or "what we do in the world."[16] Norman also
recognizes a feedback loop that guides the development of new technology upon the perception of the effects of old: beginning with goals,
cycling through intention and execution, changing our perception of the
world, leading to new goals.

Problems do not arise only when we are blinded to the world through
intoxication with our own calculative success. They tend to be much
more specifically visible. There are breaches in the cycle of invention,
which he calls gulfs of execution and gulfs of evaluation. The former
occurs when it is impossible to discover how to use a machine, because
the design does not clearly reveal its own use. A common example is
the now old-fashioned movie projector, which requires the user to
thread the film manually through an intricate series of gates and guides,
hoping for the right amount of tension between empty spaces. It is a
horrendous task, which has stymied many users and chewed to pieces
many poorly threaded films.

The projector's successor, the VCR, has circumvented this problem
by placing the threading mechanism wholly within the machine. The
videotape in its protective shell is gently eased into the housing and
mechanically set into place. Voilà! Execution is performed automatically—if all circuits are in order, all hidden, none intuitive to the touch.
Now a new gulf arises to reveal another imperfection. If the tape works,
fine. But if it doesn't work, we know nothing, as the workings are deeply
removed from our gaze and comprehension. If a film projector is failing,
we can *see* what is going wrong, which might suggest a way to fix it.
With most digital technology, the problems are located far inside the
device, and because of their electronic nature, it is impossible to perceive
what is happening. Either the "black box" sends us programmed error
messages to suggest the cause of failure (as in notoriously jam-prone

copy machines), or we trust its operation to faith or the expert. If the machine fails it is because of "gulfs of evaluation," the second category, when we are unable to discover what the problem is, and hence cannot even think how to solve it.

Such failures no longer involve the feedback of a tool's influence over time, as it reshapes our view of our own capacities. Instead, they involve the feedback of the moment—as we strike the nail, as we turn the key, press the button, *or* survey the field of predesigned artificial landscapes surrounding us. There is a sense in which our capacities and limits are challenged *every* time the instrument is used. The more the effective course of the machine is illustrated as we use it, the less explanation is needed. So for Norman, the mark of good design is a technique that reveals immediately to us what it does, how it is used, and whether it is functioning properly as we use it. This leads to a ready adoption of the new tool, and avoids misuse or inefficiency. But many modern tools are too elaborate to be so swiftly internalized.

This kind of clarity certainly makes for better-designed "everyday" objects, and as such it is a useful guide in the refinement of any new tool's particular attributes. But it may lose sight of the fact that design is more than the building of smooth-extending, easy-running gadgets to improve our lives. Winograd and Flores remark that "the most important designing is *ontological*."[17] This does not mean it necessarily creates a new world-picture or philosophy, only that we need to take heed of design's deeper, less immanent effects. Design works as a frame by bounding the range of actions of which we are capable.

Inspired in part by Alexander, Winograd's and Flores's specific task is to rethink the computer so that it will function more as a real tool and less as a model of the world. Because its workings are built out of logic, there is an immediate tendency for the universal machine to require at first a new ontology or metaphysics, one that builds its solutions entirely out of the on and the off. But this structure is an impoverished one, and to counteract it requires a critique of the philosophy behind this most subtly omniscient machine, and with that a view of technology as a whole.

Computers, according to Winograd and Flores, have generally been built upon the idea of the primacy of thought. They have been designed around logical processes which were originally meant to reflect the human mind at its most precise and least ambiguous. But when they are advanced enough to reflect an entire world built out of logic, they prove an affront to the human user, who must struggle to interact. There is always a problem of *interface*; haphazard human exploration has trouble approaching the purely logical silicon beast. It appears as an image of the logical part of us, yet in asserting independence, it tries to be more than a tool. We need techniques we can use, not mirrors that offer an image that wants to be *like* us. Computers should be "knowledge amplifiers,"[18] extending humanity in the course of activity. It is a mistake to thus sever thought from action and to consider the computer an extension only of the first. It is not foremost a tool to represent mind and world, but a device to realize our still specific intentions.

Winograd and Flores thus forecast a more modest role for this most abstract of machines. For the computer is indeed abstract, and building it according to the way we use it requires a sophisticated understanding of how humans interact with a field of pure logic, not just how problems are solved inside our minds *or* the machine's. What Winograd and Flores suggest as the direction for design really applies to all techniques where a person encounters complication: not to let the order of the device remain closed to the fact that it will be driven from without. The human mind and body must relate to it, and must be able to use it transparently, even as they are transformed through it.

So it is a mistake, then, to hope machines will become more like people. Then they will be less extending, and more adversarial. Computers should be closer to efficient and driven machines like the car or the bicycle. We do not calculate how far we need to turn the steering wheel or how hard we need to pedal to round the next bend. We just do it, feeling the results and extent of our movements. We are never simply pushing buttons or operating controls. Norman's feedback loop of tool operation is tacitly felt, and the tool does not need to talk to us.

But the more independent a tool claims to be, the greater the distance that can come between the intention and the extension. It is much harder to feel the flow of response to act, and one is more apt to feel that there is no response to one's driving actions, which are transformed into little more than motioning signals. And the more complicated and alien the device, the less one sees what has gone wrong.

For Winograd and Flores, these "moments of breakdown" contain the greatest clues to the way humanity interfaces with the other, embodied in the machine. Taking cues from Heidegger, they see the failure as a moment of releasement from transparency, when underlying mechanisms are at once unconcealed.[19] The tools themselves need to anticipate breakdowns in the course of their use, not only reporting on them, but remaining clear in cases of less than ideal operation. When the machine crashes and blithely announces "system error," how much does this tell us? Falsification is always easier than verification, as Karl Popper knew so well. More self-correction can be expected. In addition, computers, with their ability to compare and contrast large numbers of facts, may be used to anticipate breakdowns in the human world where innumerable details conflict and have to be juggled.

Any important design enacts a revision of our world, clarifying the direction of its intent. The opening of particular new possibilities always closes off others. Design will inevitably imply a blindness towards those opportunities which have been shut out. An electronic library system might allow one to call up precisely defined information with incredible speed, while discouraging the activity of browsing through the stacks where fortuitous discoveries might aid one's search. And, any library cataloging system is organized around certain units of information, while remaining clueless about others: works on the borders between established disciplines may appear in totally disparate parts of the catalog. Any system of organization is an imposition on the flow, so some kind of blindness cannot be escaped. With each improvement, we should not forget the alternatives which have been rendered invisible.

Tools create new conversations and place new connections among ideas and actions. These constitute progress only if they increase the "fit" of our intention within the larger context. The idea of fitting into the Earth as a home has already been introduced, but it may be too wide to have specific application here. An individual tool may be right when it fits into our *work*. Word processing programs that allow the writer to move blocks of text graphically without elaborate commands are sometimes said to be "smart," but these programs understand nothing of the sense of the language fragments which they manipulate. They are used because the instructions to manipulate seem transparent to the user, who may discover how to move and change blocks of text almost intuitively.

On the other hand, I write this final chapter by hand in the mountains, by a wood stove, while the rest of this text was composed directly on the machine. Surprising to me, I find the placidity of the location combines with the regular flow of the pen to enable ideas to flow smoothly onto the open page far more so than the free play of words on the screen. Tools of endless possibility will not free us from the need to act with conviction. Whenever one adopts a new technique, it is essential to remember the way things were before it came to be available.

Blindness and insight: the two prevalent consequences of directed and meticulous change. Using a tool is no simple decision to apply and to act, but a give and take that challenges goals. One message is simply always to *question technology,* to demand constant consideration of the tool by the user: "Do I really need this thing?" Better to make the questioning more precise. Ask yourself, "Do I recognize what I will gain, and what I will lose, with this technical choice?" Know also that it may not be easy to return. Practical knowledge is nothing one easily forgets; once lost, innocence is gone. Machines change faster than the moods of those engaged with them. Every technology is "a vehicle for the transformation of tradition":

> As designers and users of technology, we are always already engaged in that transformation, independent of our will. We cannot choose what the transformation will be—individuals cannot determine the course of a tradition. Our actions are the perturbations that trigger the changes, but the nature of these changes is not open to our prediction or control. We cannot even be fully aware of the transformation that is taking place: as carriers of a tradition we cannot be objective observers of it. Our continuing work toward revealing it is at the same time a source of concealment.[20]

This Heideggerian expression might sound too much as if we were *lost,* simply doomed to live through change in the throes of technology. The only answer it suggests is a vaguely heightened attention, with some greater awareness of what machines are doing to the workings of our own mind as we consider the range of our world. Of course there is nothing wrong with this, except the danger of placing too much expectation on something hardly specific. The attitude gives little advice on how to temper our interactions with machinery, or how to design better machines after the recognition of the inherent dangers. On the other hand, people continue to be constantly engaged in the use of tools, and it might be too easy to advise them just to listen and look, taking heed of what the instrument makes possible and what it seals up.

Donald Norman goes on to consider the phase of evaluation: the means by which we assess a tool as we pick it up to use it. Not content with broad generalizations such as Heidegger's jug, which is "a thing insofar as it things,"[21] he wants to investigate why precise behavior can emerge from imprecise knowledge of how a tool works. How do we figure out a TV set if all we have seen before is a radio? How do we use a saw if all we have previously known is a knife? Both natural and cultural constraints are important. Physical nature is somehow intuitive to practical human discovery. We know the sharp cuts the soft, and how intended force may be focused to make a cut. With the television we make use of cultural constraints. Used to the idea of selecting between channels, we assume the picture is tuned much like the sound. Elaborate instruction is unnecessary, even as we confront devices we have never seen before. New apparati do not challenge *all* our preconceptions of what a tool may do. The evolution is gradual, and the idea for the device must first seem reasonable and comprehensible before we attempt to use it. Tasks beyond the foreseeable range of human want are difficult to express in a tool.

This may be a superficial sense of what design is. As techniques have become more complicated and automatic, usable only after extensive specialized training, it is not always possible or desirable to get an immediate grasp on them. But in the end, once we have studied them, we would like to forget that they are intermediary and use them directly in the realization of intentions. Learning more intricate technology begins with internalizing knowledge expressed in instruction manuals or labeled parts. After we have learned to feel at home with the tool, much of the necessary knowledge will reside in our minds. As Winograd and Flores note above, we do not even think about the tool anymore when we are comfortable with it. It infiltrates our consciousness to the point of teaching us to think according to its ways. The more complicated instrument may offer more discrete options, but if we are to apply them seamlessly, we must internalize even these multiple choices. Or we do so whether we want to or not.

EXPRESSION IN CONSTRAINT

The more particular choices a technology offers, the more we learn it by apprehending its rules, not sensing its significance intuitively. The best examples of this come from the history of tools of expression, when human creativity is successively challenged by increasing *ease* of action. A paintbrush allows direct visceral contact with the material needed to mark the page. An airbrush offers precise control of color, but less character and gut variation. A computer imaging program allows the manipulation of all quantifiable aspects of the picture; programs recently introduced even imitate the specific features of various Impressionist painters. Yet the precise effect of these options on creativity is not known. More opportunity does not guarantee more originality.

The same could be said about developments in writing. The pencil connects the writer to her words through touch and sound. The pen is more permanent. The typewriter is more reliable, standard, and efficient. Can writing upon it be as personal? Never mind, we have accepted it as a fact of the manipulation of the alphabet. Now it is nearly obsolete through the convenience of the computer as processor of words. Yet with the freedom of the virtual text, as seen in chapter 4, there is little need for any version to be the final draft. Words on a screen are as flexible as the writer wants them to be. Are we well enough prepared to enjoy this freedom?

The examples could go on. My favorite is from the evolution of musical wind instruments.[22] The first flute was only a tube across which one could blow, playing a series of notes off the Pythagorean harmonic series. There is only Klee's natural order, not the artificial. To change the pitch, you can blow harder or softer, that is all. There are no holes for the fingers to cover, no musical extension of the hand. (This kind of instrument survives in the *selje flute* of Norwegian folk music.)

Then comes the technically primitive Japanese *shakuhachi* flute, which offers five holes and suggests music based around a pentatonic scale. But the mouthpiece is flexible enough to allow a lot of direct expression in the embouchure, bending the pitch with changes in angle, changing the tone with the shape of the lips. All variation comes from *within* the human player, so that each person plays quite differently, producing a personal sound, as so many organic and physical constraints are combined in a different way each time the flute is blown.

Then six hole flutes become common, in line with the diatonic scale. Pitches are more fixed, and there are more discrete choices. Music of six and seven note scales becomes easier to play. Note that it is still possible to play notes in between those suggested by the fingering holes. It requires more intuition and practice from within the player again, with less specific information contained in the instrument. A more precisely chromatic music demands a scale of twelve notes, which leads to chromatic keys added to the basic forms. Now players can slide easier up and down a scale of finer, definite intervals, each exact and uniformly set out, as on a clarinet. This is Klee's artificial order, rather than a continuous, intuitive cry or glissando like the original whistle of the wind or the siren of a loon.

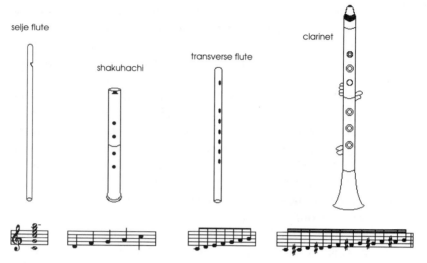

Figure 9. Increasing fingering possibilities increases the ease of playing

More complicated musical instruments promise a flurry of more exactly set notes, with maybe less immediate, fluid feeling. More rules have been installed between the player and the played. Music composed of many fast notes is "easier" to play, but the unquantifiable, nameless quality is still elusive, if clouded by so many established choices.

Recent developments have put the distance of computer technology in the scheme. It is now possible to make woodwind sounds with a digital input controller, called a wind synthesizer. The device looks like a wind instrument, but it makes no sound. All that is sent out are digital signals, to be interpreted by a synthesizer and reconfigured as timbre. One plays the instrument and hears sound, imagining causality. Yet there is no direct link that allows us to feel the producing of the music itself. All is calculated, all measured and re-created. As we play the digital horn, we need not grasp how it works; it is successful if we feel as if we are playing a natural wind instrument, with all the expression, all the intuitive take on the natural laws of physics. But an abstract, black box is making the sound. We believe it only when caught in an illusion.

If this connection works, it is the result of programming, a bond quantified if we are able to bridge the gap of digital distance that a "thinking" machine implies. We will have calculated, we will have explained. The skill of shaping the sound, which usually takes years of training, will have been programmed into the interface between person

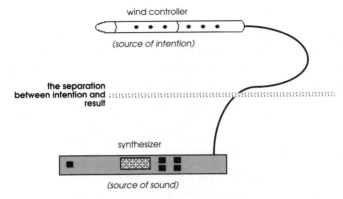

wind controller

(source of intention)

the separation
between intention and
result

synthesizer

(source of sound)

Figure 10. The black box musical model separating controller
from sound source

and machine, filling the distance with the illusion of transparent expression of a human idea through the tool into the world. Yet this electronic instrument uses a method of input with which wind players are familiar: the blend of the analog control of breath with the digital control of covering successive holes and keys to change the pitch, held over from the kind of machine that let us directly take hold of physical properties. Thus we *crave* the laws of nature, even when technology seems to liberate us from them. The habit of command remains, freed from effect.

Here is a technology symbolically opposite the tunnel that takes us to the eternal reaches of nature. The eloquent instrument is a vehicle for human artistic statement. We possess at first our own idea and feeling, and the tool is the thing that releases this idea to the world in physical form. The flute extends us, but toward what? The tunnel brings us into nature, but what is the "us" which is brought? And when one choice is made, do not forget what it denies. Look ahead, behind, and across the long frontier of technical possibility.

The insights offered by a technique tend to close off those options which the device is unable to deal with. As each machine suggests more problems to be solved or features to be improved, those questions less susceptible to a technical answer become impossible to see.

HUMANITY EXTENDED

A whole slew of divisive dichotomies has now been introduced, to show how technology propels its authors toward a goal, while ensuring that its own rules will continue to be followed. There is always a danger with such category-making: the temptation of deciding that one side is preferable to the other, which is determined inferior. And there is also the danger of considering all the polarities as referents to the same thing, assuming that the problem has only two sides. These are mistakes— divisions do not end the flow of world between bow and lyre. Each distinction is momentary, and separate. Each makes the problem look different, and each needs to be taken in itself as a guide to help form a balance between the two choices. Do not imagine the world holds to duality! As Emily Dickinson observes,

> This World is not Conclusion.
> A Species stands beyond—
> Invisible, as Music—
> But positive, as Sound—
> It beckons, and it baffles—
> Philosophies—don't know
> And through a Riddle, at the last—
> Sagacity, must go—[23]

Wisdom must pass through something to change things, and the riddle might be technology, the practical engagement of humankind with the world. There is no final arbiter, but the list of divisions is a kind of elevated beginning, starting the observer "a few feet above the ground." The invisible goal beyond remains impossible to see, but perhaps it can be heard.

Figure 11. Various dichotomies through which to
consider technology

These may be divisions, but they are all inseparable parts of the whole technological enterprise. I present them in this manner to avoid identifying half with the left column, half with the right. Like Spinoza's creative *naturans* and manifest *naturata*, each a part of *Natura*, the one, all these demonstrable perspectives are an inevitable part of *techne*, the irrefutable medium for human engagement with the world.

If we still want to hold onto nature as a goal for the extension of humanity, it must not only be the ground of irreversible forces, but also a surrounding human sense that will come near as an achievement. This I have described as fitting-in, shaping the world as a home, or building a place in line with the world; not by mastering it, but by seeing our decided range—to explore, to learn, to marvel at, to enjoy, but not to destroy or eradicate.

It is common these days to make this claim more specifically modest by asking for *sustainable* technology: a kind of machinery that will be workable and not detrimental to the Earth long into the future. The problem is that technologies change so fast these days, with humanity and its context in a mutual tumult that exceeds our ability to keep track of them. To use machines as if we wish them to be used forever neglects the fact that devices seem to be stages in the development of something beyond themselves. Yet "improvements" are often not directed toward any goal apart from the streamlining of technical processes already in place. They rarely question the goal of the technique itself. Once refinement has begun, it is hard to stop.

The movement to temper technology according to nontechnical forces such as natural limits or overriding cultural constraints has been summarized in various ways. Consider the slogans "Appropriate technology!" "Small is beautiful!" "Intermediate technology!" "Right livelihood!" "Advanced technology is only that which advances the goal of each culture!" There is nothing wrong with these affirmations as such. The question immediately becomes one of scale—by whose and what criteria are we to limit the kind of technical progress which seems to build on itself, revealing new problems and new directions with each new solution?

The opinion that the simpler might be better can be traced at least as far back as Lao Tzu, and his opinion that humanity should hold back ability in order to best apprehend what surrounds us. Here is the kind of homeland he believes we should wish for:

> The ideal land is small
> Its people very few,

Where tools abound
Ten times or yet
A hundred-fold
Beyond their use. . . .

The folk returns
To use again
The knotted cords.
Their meat is sweet;
Their clothes adorned,
Their homes at peace,
Their customs charm.[24]

Happy in the restrained, balanced world, people will not want to travel even to the nearest alien country, because they have all they want within their own borders, with enough kinds of tools for the inhabitants to say yes to some, no to others.

But this intimate life is not necessarily complete—else why would we have left it far enough behind (even in Lao Tzu's day) to long for it nostalgically? Cities are built, their walls sealed, culture ensconced in conversation. Explorations continue to bring word back to the center. Empires are established, far-off nations made captive. And why couldn't humanity take to the heavens and leave this overused planet behind, once we all, like former U.S. Secretary of the Interior James Watt, become "bored with the Grand Canyon"?

The "appropriate" reaction is not necessarily to recoil from the gigantic; nor should we accept expansion blindly. The range of our civilization has grown tremendously, but in omnipresence it has made the world seem small by comparison: included, explained. Extension makes even the most distant reaches run along human criteria, and the remote whorls of the galaxy seem "intimate" when explained by our theories and conjectures. The right, or chosen tools must not just hold back before obliterating a nature which they have simplified in order to conquer. They must also succeed at enriching the life of human beings together, helping us realize the way we increase ourselves by making and sharing things together. Marx described the arena of humane production as a domain of many mirrors, each of us reflecting our authentic, appropriate essence upon all others. It is a happy idyll, where we use tools to extend ourselves toward each other as much as to the world. Technology is never only about the individual trying to decide what tool to pick up and how to make use of it. It is also concerned with binding us within social structures, organized but hopefully not dehumanizing.

Ivan Illich picks the term *conviviality* to express the kind of making that enables people to enjoy life with each other. Here hopeful society seems the at-hand limit, not the inaccessible mystery of fragile natural forces. Tools are as much instruments of social relationship as they are rafts to ride the currents of nature. We cannot afford to give in to them, while remaining prepared to accept what they do to us. Our self-image cannot be so closed as to resist enhancement via technique and order, yet it should never be so footloose as to be obsessed with the machine, forgetting its reason:

> A tool can grow out of human control, first to become our master, and finally our executioner. Tools can rule us sooner than we expect: the plow makes us lord of a garden but also refugee from the dust bowl. *Nature's revenge* can produce children less fit for life than their parents, and born into a world less fit for them. *Homo faber* can be turned into sorcerer's apprentice. Specialization can make our every day so complicated that it becomes estranged from our activity. Addiction to progress can enslave all humanity to a race in which none ever reaches the goal.[25]

Beware, but do not turn away from the sense of motion that is central to so many current attempts to define the human. Technical change cannot be separated from our essence. The humanity which is extended is already a technical creation. But get a hold of it, see it before it sees you. Direct change, do not defy it. Technologies should continue to extend us, not replace us.

This in the end is the best reason to hold onto the image of technology as human extension, spreading still farther outward into the world through the application of tools. Yet what are we expanded toward? And just what *is* extended, if human life is so transformed by the engagement and involvement with technology? It is possible to consider *all* human activity or thought to be *techne,* but only to the extent that it alters the world and ourselves upon a claim of constructed, rather than discovered, order.

Inside us is a human nature that wants to explain, inhabit, and map out the universe. Out there is a determinate web of forces which binds our behavior, but whose subtle variation and completeness will still elude our methods. The more we are extended, the more it is necessary to become attuned to those parts of the world which we have come to depend on and to use. We should never simply extract what we need from the Earth, forgetting to admire what it endlessly offers, far beyond the extent of Bacon's conceivable utility. The complication that arises through modern technology bars us to the complexities of nature released. The tremendous dynamism implied by modern technology obliges us to pay attention to more interconnections in the world than we ever thought possible, important, or so delicate. As we make use of more, we do not control more, but rely on more.

Depending on nature, do we at last become a part of it? Only if we can reconcile our increasing transformation with the desire to find a home. We have to want it—this must also be intended. Here is a key to the enigma of Pascal that began this work: "Nature diversifies and imitates, artifice imitates and diversifies." What is the difference between technology and nature? Our intentions are replicated, and then one becomes many. In the external world we choose whether to see parts or wholes, once divided, that strive to hold together. There is no purpose out there we can see, for we would have to know much more about the world to be able to be sure if there is such a thing as design or direction apart from the responses we observe to our own actions. Only in this sense do we never learn more than ourselves. Confident of what we have shown to cause and effect, we find it easy to point purely in controlled directions—this is technology evolving upon itself. But there is a *right* speed to proceed at, and this is the one that advances with necessary caution, asking why a process should be sped up, simplified or made more efficient. Technical advance in itself is no certain end for humanity. Or, to quote Pascal again, moving "too fast *or* too slow, *we* are unintelligible," and with us, all that we do, all that we make, and all that we learn to see or imagine.

The second epigraph? Elias Canetti depicts the stalwart power of the hand that begins the reach toward technology. Words and objects alike come from the clasp of the hands. The ordered world is not at first seen, but felt. The habit of the hands must be well in place before the mind rises to extend motion through a device. The place entwined with the fingers then has only the limits of human ingenuity to face.

What of the suggested arrogance of having humanity appear so strong? After reading this far, it may seem to you that I have presented a highly optimistic vision of technology that puts tremendous faith in human ability to check our own long history of continued expansion. My position might seem, then, to possess the kind of "arrogance of humanism" so aptly attacked by biologist David Ehrenfeld. But by humanism he intends a "supreme faith in human reason," denying existence of God or the supernatural, believing that the "undirected powers of nature in league with blind chance"[26] are things that we can and will master, in the pursuit of avowedly human purpose. Thus defined, humanism is bound to have a certain prejudice against the natural, nonhuman world.

My answer is that we are humanity, and thus any of our ideas and actions may be called human. We shape and are limited by a context which may be called "nature," whose appearance changes as we enforce our place within it. There will always be much that we do not know about this world that encompasses and unfolds us. No success of technique should deny this mystery, and many tools serve directly to reveal it. Thus it is misleading to oppose technology to any Heideggerian "openness" to the wonder that we are thrown into the world. Technology may lure us into concentration on its own furtherance, forgetting larger, ancillary spheres, but it does not require such single-mindedness. Oppositions like "human" vs. "natural" do not help in this regard. Perhaps we should do away with these words altogether.

But they are all that our language now has, and one needs to respect language as the oldest, most evolutionarily weathered, and least planned of enduring, sustainable technologies. There must be something to these distinctions if they have lasted so long. Maybe it is time to reform our language so that we do not talk as if we were set in contest against the Earth. I hope I am not making a compromise in my acceptance of technology as a route towards nature rather than a black box that foils and regiments our direct apprehension of the beyond. The only nature we know is a realm reconfigured as our tools construe it, even to the point of *seeming* so independent of us that it emerges as a wonderful other, which we believe we can touch only if we release ourselves from all devices. But that vision of nature is just one reaction to civilization, and far too simple to encompass the idea of nature as the ground that makes human development possible *and* the goal of an enduring home that awaits.

Ehrenfeld is right to be worried that excessive faith in technology is a human arrogance, more accurately a blindness to the effects of our actions, which may lead to self- and utter destruction. But only further extension of humanity, not reaction against it, will save us. The arrogance is of exclusively human interest—not of humanity itself. If we begin to examine how our own interests are tied to the world's, we will not need to encourage the dichotomy that separates us from our nature as cause and result.

Among those human abilities that encourage Ehrenfeld, giving him some hope that we will pull ourselves and the world together, is the "capacity for men and women to stand alone, triumphant, in simplicity, independent of the constructions of devices of society and the plans of other people."[27] We can step back from all machines and measures; we can gaze at the mountains and try to forget all it took to get us to this viewpoint. A simple essence: we can reflect. Something in us remains untouched by techniques. It is this part that has not changed from the times of Heraclitus and Spinoza to our own. Is that humanity? No, just a part of it. What we mustn't do is oppose this part to the other, technical part. Both flow together. As we gaze at the distant clouded horizons, think not of humankind before the fall, but of all the mistakes and triumphs of civilizations and cultures that now permit us to care. Turn around and make sure that the range of experience and possibility will not be broken.

The world is always much more than we can ever know of it. But the truth of the world is not the question of this work. The concern here is the way we inhabit the earth, and how this affects what we see. Even human *survival* is not at issue; that is one step down from life. To keep hold of the spread of the living, we should not oppose the animate to the inanimate. The things we build expand what we are. But where do they take us?

In what direction does the expansion go? All around, in every human dimension, making our capacities larger than unadorned life. I have argued that the machine leads us toward the limits of nature, so that, as we establish ourselves through and with technology, we are brought closer to a practical understanding of whatever remains beyond the human. We identify with the world not as we look at it from a detached vantage, but as we use and are tested by it. If we pay attention only to the mechanisms of the devices we construct—how to improve them, how to build worlds upon them alone—we forget the context that inspires and limits all tools. This is the circular frame of our original desires and the guiding effects of the larger world, which becomes more bound to us the more we test its balance. We demand a place as part of nature because this demand accepts the Earth as our home. We will never learn all of it, but the modifications we make on it should share our place within it, rather than plow our ideas over all in their reach.

But what we use changes what we see! From the earliest and clearest tools, we wish the world would work like what we make. So we make it appear so, and we look only at certain things. The bow and lyre change our vision of Earth as much as the computer—even if the latter claims to be the very structure of reality rather than merely one experience of it. Nature, as it surrounds us, eludes the categories by which we judge it.

If you turn to look away from tools, something shines directly through. Up here in the mountains it has been mostly gray, socked in, empty of referent, a fine environment for nurturing abstract ideas. But just yesterday the clouds lifted, and the distant ranges once more became clear. The evening light bathed the soft shapes orange, and I looked at them, yearning to be up there inside. Not to sail through them in a machine, but to be as they are—solid, yet melting in air. I tried to test my gaze. Clouds, orange—could I even conceive of such things apart from imagining how they might be *techne,* things that are made? Why do this? *Technology need not overtake wonder at the world.* It must serve a part of that awe, not subvert it and dream order over all. We can still respect our own prowess *unless* it overrides the Earth.

Then, after the clouds, a tiny piece of rainbow on the horizon. I will never see such in the sky and forget what I know about the bending of colors in white light, something that a glass tool called the prism has taught us. But we have not made the light. Precisely because *we* have *not* made it does it shine so solidly in the mind. For things do happen in the world beyond human intent. And this will not be explained. *Outside the walls, machines and designs—the river, the trees, the birds.* Humanity may be extended toward the world, but this brings us the challenge of stepping out into the direction of the unknown, learning to inhabit it before we end it.

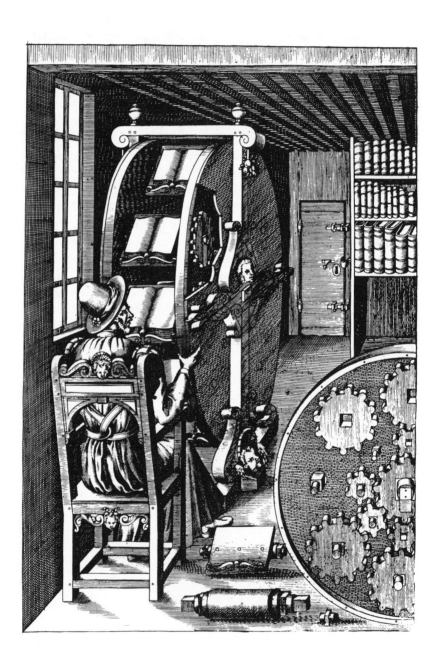

Notes

In citations from classic philosophical works, the conventional section and line numbers are given, which differ from Plato to Aristotle to Spinoza. Page number references are given to current and available editions, such as Plato's *Collected Dialogues,* edited by Huntington and Cairns; *Introduction to Aristotle,* edited by Richard McKeon; and the William Hale White translation of Spinoza's *Ethics.*

The symbol ◊ indicates that sexist language has been altered in the quotation to allow references to "man" to apply to "humanity" in deference to present readers. Is this fair to history, or pandering to political correctness? When translating from other, less sexist languages to English, it is only the latest case of modernizing an original to get its message across to an evolving public. If the original was English, then it is a touchy kind of editorializing, but I am hoping it will enable those offended by the masculine bias of our language to ponder these messages with greater seriousness.

PREFACE: LATENT LANGUAGE

1. Martha Teach Gnudi, "Biographical Study of the Author," *The Various and Ingenious Machines of Agostino Ramelli,* p. 13.

2. Agostino Ramelli, "Preface: On the Excellence of Mathematics," *The Various and Ingenious Machines of Agostino Ramelli,* p. 47. ◊

3. For the results of this work, see Arne Naess with David Rothenberg, *Ecology, Community, and Lifestyle.* See Warwick Fox's *Toward*

a Transpersonal Ecology for an analytic overview of the whole field of deep ecology.

4. Theodor Adorno, *Aesthetic Theory,* p. 89.

CHAPTER 1: UNEXPECTED GUILE

1. Translation by Jonathan Barnes in *Early Greek Philosophy,* p. 101.

2. Italics within quotations will generally indicate *my* emphasis.

3. Heraclitus, Fragment B51; *Early Greek Philosophy,* p. 102.

4. Octavio Paz, *The Bow and the Lyre,* p. 262.

5. Lewis Mumford, *The Myth of the Machine,* p. 115.

6. Plato, *Euthyphro,* 13e–14b; *Collected Dialogues,* p. 183.

7. Plato, *Philebus,* 55e; *Collected Dialogues,* p. 1137.

8. Ibid., 56a; *Collected Dialogues,* p. 1137.

9. Ibid., 58c; *Collected Dialogues,* p. 1140.

10. Aristotle, *Nichomachean Ethics,* VI, 4, 1140a; *Introduction to Aristotle,* p. 427.

11. Plato, *Meno,* 97d; *Collected Dialogues,* p. 381.

12. Aristotle, *Physics,* II, 2, 194b; *Introduction to Aristotle,* p. 121.

13. Aristotle, *Physics,* II, 8, 199a–b; *Introduction to Aristotle,* p. 134–135.

14. Aristotle, *Metaphysics,* I, 1, 981a; *Introduction to Aristotle,* p. 244. ◊

15. Dewey, *Experience and Nature,* p. 21.

16. Ibid., p. 20.

17. Ibid., p. 86.

18. Ibid., p. 91.

19. Ibid., p. 103.

20. Ibid., p. 282.

21. It is certainly possible to find the roots of the cyclical vision of technology which pervades this work within Dewey's writings, but one would need to transgress the limits of his instrumentality to do so. I find within his writings many perceptive comments on how tools lead us on toward ends, but the circle is not there, as he fails to realize how much his own intentions are affected by an intoxication with all things technical. See Larry Hickman's *John Dewey's Pragmatic Technology* for more detail.

22. The conventional (Macquarrie and Robinson) translation of these two terms are, respectively, present-at-hand and ready-to-hand, which fit into their staggering attempt to recreate a Heideggerian language within English. In my effort to make use of these concepts, I am

trying to use more conventional discourse, so I have decided to discuss them without introducing too much distracting jargon.

23. Martin Heidegger, *Being and Time*, I, 3, A, ¶15, p. 98.

24. Ibid., p. 101.

25. John Dewey, *Experience and Nature*, p. 60.

26. Joseph Weizenbaum, *Computer Power and Human Reason*, p. 9. ◊

27. Philosophical application of terminology always risks comparison with common usage of the terms it employs. When one is concerned with sharpening the edges of concepts so that they will be better understood by all, it is appropriate to remember the inherent fluidity of the meaning of words within natural languages, and not to supplant this flexibility with overdetermination. Hence, definitions should not be too exclusive or narrow in philosophy, else it will be impossible to write clearly on general, foundational topics in an interesting way. This has led some philosophers to forego such an enterprise, but I think that just begs the question. Definitions in this work will, as in the most durable of philosophies, suggest clarified uses of familiar concepts without forgetting their original ambiguity.

28. Is a machine the same as a tool? As we use it, yes. As metaphors, machines suggest some autonomous intricacy, something more self-contained than a hammer. Ramelli has drawn "ingenious machines," true, but when it comes time to piloting them, they too serve as tools. At this point I wish to emphasize the similarities, not the differences.

29. For a thoroughly eccentric look at how the telephone underlies all of modern communication and philosophy, see Avital Ronell, *The Telephone Book*.

30. Sophocles, *Antigone*, 365–366, p. 342. Literally: "having, in the inventiveness of *techne,* something cunning beyond expectation." My translation.

31. Edmund Husserl, *The Crisis of European Sciences and Transcendental Phenomenology*, II, §9, a, p. 25. ◊

32. Ibid., II, §9, g, p. 48.

33. Ibid., II, §9, h, p. 49. See also Husserl's essay on "The Origin of Geometry" (reprinted in *The Crisis*, pp. 353–378), an attempt to suggest what fundamental characteristics our lifeworld shares with that of the first geometers, outlining our progress in reason as an attempt to get a grip on this one common world which all humanity shares.

34. The German philosopher of technology Friedrich Dessauer was puzzled by the recurrent testimony of inventors that they felt their constructions were *discovered* rather than independently invented according to individual whim:

The inventor does not view what has been gained from his creation with the feeling "I have made you" but, rather, with an "I have found you. You were already somewhere, and I had to seek you out for a long time. If I could have made you out of myself alone, then why would you have concealed yourself from me for decades—you, an object found at last?" (Friedrich Dessauer, "Technology in its Proper Sphere," in *Philosophy and Technology,* ed. Carl Mitcham and Robert Mackey, p. 323.)

The "found object," for Dessauer, comes from an *a priori* realm of possible inventions, accessible to us through a fourth faculty which is meant to be added to Kant's original three faculties of pure reason, practical reason, and judgment. To explain the perplexing observation that the principles behind tools seem to be discovered rather than fabricated, he locates them in their own preexisting conceptual space, which humans successively penetrate and explore. This is a wild idea, but certainly in the venerable spirit of idealist philosophy, looking within to explain what is outside us, a task perhaps inspired by recoil from a rapidly changing external world. Does technology as a whole need an independent conceptual place of its own? Though I think the idea is something worth putting forth, it fails because technology cannot exist as a set of completed concepts before they are ever thought of or applied. The metamorphosis that occurs in practice is far too great for that. Still, it is important to consider the fact that the most important of inventions are usually conceived by their proponents as discoveries, so they do extract something out of an external reality by finding new ways in which humanity can extend into this reality.

As none of Dessauer's larger works have been translated into English, the summary of his views in the introduction to the cited volume (pp. 21–24) and in Egbert Schuurman's *Technology and the Future* (pp. 102–107) may also be of interest.

35. George Basalla, *The Evolution of Technology,* p. 2.

36. Ibid., p. 218.

37. Blaise Pascal, *Pensées,* XV, 199, "Man's Disproportion," p. 90.

38. Paul Levinson, "Information Technologies as Vehicles of Evolution," p. 32.

39. Ibid., p. 34.

CHAPTER 2: EXTENSION'S ORDER

1. Classifications of the whole of technology are surprisingly hard to come by. It is more common to find classification systems for mechanisms alone. These are descended from the Linnaean approach devised for the natural world, suggesting that those who tried to sort out the

different types of machines in the eighteenth and nineteenth centuries understood technology in evolutionary terms. But machinery was changing in too many ways at once, and it proved impossible to contain within any list of possible motions or combinations of motions. See David Channell, *The Vital Machine* (pp. 80–81) for a survey of classification attempts, none of which tries to include as much as I have here under the rubric of technology.

2. See S. Allen Counter, *North Pole Legacy*.

3. Paul Levinson, "Information Technologies as Vehicles of Evolution," p. 36.

4. Ibid., p. 42.

5. Jim Harrison, "Unimaginable Visions," p. 28.

6. See Abraham Maslow, *Towards a Psychology of Being*, for an account of the steps toward self-realization.

7. See Arne Naess, *Ecology, Community and Lifestyle*, chapters 3 and 4.

CHAPTER 3: NATURE AS CONTEXT

1. See Jens Allwood, "Natural Resources: Language, Beliefs, and Concepts," for a summary of these many meanings.

2. Wolfgang Schadewaldt, "The Concepts of Nature and Technique According to the Greeks," p. 160.

3. Aristotle, *Physics,* II, 192b.

4. Here I will capitalize "Nature" when referring specifically to Spinoza's encompassing use of the term.

5. Baruch Spinoza, *Ethics*, 1, Proposition XXIX, p. 65.

6. Ibid., 1, Definition VII, p. 41.

7. Ibid., 1, Proposition XXXI, p. 66.

8. Ibid., 2, Definition VI, p. 80.

9. Ibid., 2, Proposition VII, p. 83.

10. Ibid., 2, Proposition XL, p. 110.

11. Ibid., 4, Proposition II, p. 192.

12. Baruch Spinoza, "Letter XXXII," *The Correspondence of Spinoza*, p. 209.

13. Baruch Spinoza, *Ethics*, 5, Proposition XXIV, p. 269.

14. Ibid., I, Definition V, p. 41.

15. Ibid., 4, Preface, p. 188.

16. Ibid., p. 188.

17. Juan Luis Vives, *De causis corruptarum artium,* quoted in Paolo Rossi, *Philosophy, Technology, and the Arts in the Early Modern Era,* p. 6.

18. Francis Bacon, *The New Organon,* I, LXXIV, p. 72.

19. Ibid., I, CX, p. 102.

20. Francis Bacon, "De augmentis," *The Works of Francis Bacon,* pp. 496–497.

21. "New Atlantis," in *Selected Writings of Francis Bacon,* pp. 574–581.

22. Ibid., p. 583.

23. Paolo Rossi, *Philosophy, Technology, and the Arts in the Early Modern Era,* p. 160. The preceding pages which lead up to this reading discuss possible interpretations of the following direct excerpt from Bacon's *Novum Organon,* I, CXXIV:

> Atque ipsissimae res sunt, in hoc genere, veritas et utilitas: atque opera ipsa pluris facienda sunt, quatenus sunt veritatis pignora, quam propter vitae commoda.

In Rossi's literal translation:

> In this respect things themselves, as such, are at one and the same time truth and utility, and works themselves must be esteemed more as pledges of truth than that they be pursued because of the comforts of life they afford.

The commonly accepted translation by Spedding is:

> Truth, therefore, and utility are here the very same things; and works themselves are of greater value as pledges of truth than as contributing to the comforts of life.

Spedding recognized an anomaly in his reading of the Latin *ipsissimae res,* but future commentators tended to accept the translation as gospel, clouding a more subtle ambiguity that may have been intended.

24. Karl Marx, "Economic and Philosophical Manuscripts of 1844," in *Karl Marx: Early Texts,* p. 167. ◊

25. See Yermiyahu Yovel, *Spinoza and Other Heretics,* Vol. II: *The Adventures of Immanence,* p. 86.

26. Karl Marx, "Economic and Philosophical Manuscripts of 1844," in *Karl Marx: Early Texts,* p. 135.

27. This view is developed by Yovel, *The Adventures of Immanence,* p. 96.

28. This and all the following Marx quotes come from the final pages of "On James Mill," in *Karl Marx: Early Texts,* pp. 201–203.

29. See "Letter XXXII," *The Correspondence of Spinoza,* pp. 209–211, which demonstrates how all objects we encounter are conceivable as both parts and as wholes.

30. Martin Heidegger, "Building Dwelling Thinking," in *Poetry, Language, Thought,* p. 150.

31. Martin Heidegger, "The Question Concerning Technology," in *The Question Concerning Technology and other Essays,* p. 16.

32. Ibid., p. 27.

33. Ibid., p. 28.

34. Ibid., p. 33.

35. Martin Heidegger, "Language," in *Poetry, Language, Thought,* p. 190ff.

36. Martin Heidegger, "Poetically Man Dwells," in *Poetry, Language, Thought,* p. 227.

37. Martin Heidegger, "The Question Concerning Technology," p. 10.

38. Lewis Mumford, *Technics and Civilization,* p. 319. ◊ Mumford is quoting his teacher Patrick Geddes. For more information on Geddes, an immensely creative individual in urban planning, peace studies, international education, and human ecology, see Philip Boardman, *The Worlds of Patrick Geddes.*

39. Ibid., p. 324.

40. Ibid., p. 329. ◊

41. Ibid., p. 363.

42. Ibid., p. 371.

43. Ibid., p. 372, plate 3.

44. Ibid., pp. 378–379.

45. Marshall McLuhan, *Understanding Media,* p. 8.

46. Marshall McLuhan and Quentin Fiore, *The Medium is the Massage,* p. 145.

47. Marshall McLuhan, *Understanding Media,* p. 349.

48. Ibid., p. 353.

49. Marshall McLuhan and Quentin Fiore, *The Medium is the Massage,* p. 63.

50. For a compendium of reviews and critiques of McLuhan, see *The McLuhan Explosion,* ed. Harry Crosby and George Bond.

51. See Richard Schickel, "Marshall McLuhan: Canada's Intellectual Comet," *Harper's Magazine* (November 1965), reprinted in *The McLuhan Explosion,* p. 90.

52. Yet it is no coincidence that the late John Cage, one of the prophets of chance operations as a tool in art, was also an avid devotee of nature, and has said he would prefer mushrooms to music if he had to make the choice. See his *Silence.*

53. Hans Magnus Enzensberger, "Constituents for a Theory of the Media," in *The Consciousness Industry,* p. 113.

54. Lewis Mumford, "An Appraisal of Lewis Mumford's *Technics and Civilization* (1934)," p. 534.

55. Lewis Mumford, *The Pentagon of Power,* pp. 293–294. ◊

56. Ibid., p. 297.
57. Ibid., p. 298.
58. Ibid., plate 17, following p. 340.
59. Ibid., pp. 395–396.
60. See Sigmund Kvaløy, "Complexity and Time: Breaking the Pyramid's Reign," in *Wisdom and the Open Air,* pp. 113–152.

CHAPTER 4: NATURE IS MADE

1. Hans Magnus Enzensberger, "E. J. M. (1830–1904)," in *Mausoleum: Thirty-seven Ballads from the History of Progress,* p. 113.
2. Plato, *Republic,* Book X, 616c; *Collected Dialogues,* p. 840.
3. Plato, *Timaeus,* 33b; *Collected Dialogues,* p. 1164.
4. Hero of Alexandria, *Pneumatics,* passim.
5. See Derek de Solla, "An Ancient Greek Computer," quoted in J. David Bolter, *Turing's Man,* p. 20.
6. Lewis Mumford, *Technics and Civilization,* p. 13.
7. Nicole Oresme, *Le Livre du ciel et du monde,* p. 289; quoted in Bolter, *Turing's Man,* p. 27. ◊
8. René Descartes, *Principles of Philosophy,* II, 4; *The Essential Descartes,* p. 338.
9. Ibid., II, 25; p. 346.
10. René Descartes, *Discourse on the Method,* Part V; *The Essential Descartes,* pp. 138–139.
11. Julien Offray de la Mettrie, *Man a Machine,* pp. 131–132.
12. Ibid., p. 128.
13. Ibid., p. 105.
14. Joseph Needham, *Man a Machine,* pp. 86–87. Needham's book was written "in answer to a romantical and unscientific treatise written by Sig. Eugenio Rignano and entitled *Man not a Machine."* Rignano's book, subtitled "A Study of the Finalistic Aspects of Life," was published by Kegan Paul in the Psyche Miniature Series in 1926.
15. Ibid., p. 101.
16. René Descartes, *Principles,* 37; *Philosophical Writings,* p. 203.
17. Quoted in Abbot Payson Usher, *A History of Mechanical Inventions,* p. 71.
18. Ibid., p. 355.
19. Norbert Weiner, *Cybernetics,* pp. 41–42.
20. Quoted in Cecelia Tichi, *Shifting Gears,* p. 36.
21. Quoted in Tichi, pp. 39–40.
22. Norbert Weiner, *Cybernetics,* p. 44.
23. See Hans-Georg Gadamer, *Truth and Method,* passim.
24. See Ludwig von Bertalanffy, *General Systems Theory,* p. 43.

25. Fritjof Capra, quoted in Warwick Fox, *Toward a Transpersonal Ecology*, p. 170.

26. von Bertalanffy, *General Systems Theory*, p. 53.

27. Ludwig Wittgenstein, *Tractatus Logico-Philosophicus*, proposition 7, p. 74.

28. Quoted in Pamela McCorduck, *Machines Who Think*, p. 27.

29. Usually the test is described as using only two symbols, the zero and the one, but the machine was hypothesized around any finite set.

30. He never made the claim himself, and others have tried to take credit for themselves. It is clear, though, that von Neumann made comparisons as early as 1945 between the machine he was proposing and the human nervous system. See Pamela McCorduck, *Machines Who Think*, p. 62.

31. See Alan Turing, "Computing Machinery and Intelligence," pp. 434–460. Turing introduced his ideas on artificial intelligence as early as 1947 in a paper entitled "Intelligent Machinery," only published twenty years later in *Machine Intelligence 5*, ed. B. Meltzer and D. Michie.

32. Ibid., p. 452.

33. Ibid., p. 442.

34. According to Joseph Weizenbaum, *Computer Power and Human Reason*, p. 3.

35. Ibid., p. 4.

36. K. M. Colby, J. B. Watt, and J. P. Gilbert, "A Computer Method of Psychotherapy: Preliminary Communication," quoted in Weizenbaum, *Computer Power and Human Reason*, p. 5.

37. Quoted in Weizenbaum, *Computer Power and Human Reason*, p. 6. ◊

38. An engaging report of the initial controversy over Weizenbaum's critique may be found in McCorduck, *Machines Who Think*, pp. 305–328.

39. Ibid., p. 101.

40. Ibid., p. 199.

41. See Roger Penrose, *The Emperor's New Mind*, p. 13.

42. Quoted in the *New York Times Magazine*, 14 January 1990, p. 65.

43. McCorduck, *Machines Who Think*, p. 352.

44. Sherry Turkle, *The Second Self*, p. 315.

45. Ibid., p. 316.

46. Ibid., pp. 294–295.

47. Penrose, *The Emperor's New Mind*, p. 447.

48. Doug Engelbart, "A Conceptual Framework for the Augmentation of Man's Intellect," quoted in Howard Rheingold, *Tools for Thought*, p. 181.

49. Ibid., p. 139.
50. Jaron Lanier, "Lucy in the Sky with Computers," p. 7.
51. See Gibson's contribution to *Cyberspace: First Steps*, ed. Michael Benedikt.
52. Donna Haraway, "A Cyborg Manifesto," *Simians, Cyborgs, and Women*, p. 181.
53. Jaron Lanier, "Lucy in the Sky with Computers," p. 8.

CHAPTER 5: BEFORE THE END

1. Karl Jaspers, *The Atom Bomb and the Future of Man*, p. 194. ◊
2. Niccolò Machiavelli, *The Prince*, p. 89.
3. Lao Tzu, *Tao te Ching*, pp. 110 and 83.
4. Quoted in Richard Rhodes, *The Making of the Atomic Bomb*, p. 676.
5. Ibid., p. 672.
6. Albert Einstein, fund-raising telegram for the Emergency Committee of Atomic Scientists, 23 May 1946. Quoted in Steven Hilgartner, et al., *Nukespeak*, p. 1.
7. Henry Stimson, quoted in Richard Rhodes, *The Making of the Atom Bomb*, p. 642. ◊
8. J. Robert Oppenheimer, speech to the Association of Los Alamos Scientists, November 1945, ibid., p. 761.
9. Edward Teller, on rejecting Leo Szilard's boycott of the use of the bomb, ibid., p. 770.
10. Niels Bohr, ibid., pp. 532–534.
11. Quoted in A. G. Mojtabai, *Blessèd Assurance*, pp. 173–174.
12. Robert Jay Lifton, *The Future of Immortality*, p. 155.
13. Ibid., p. 125.
14. Ibid., p. 274.
15. Hans Jonas, *The Imperative of Responsibility*, p. 27.
16. Spencer R. Weart, *Nuclear Fear*, p. 430.
17. Ibid., pp. 187–188.
18. Francesca Lyman et al., *The Greenhouse Trap*, p. 9.
19. Bill McKibben, *The End of Nature*.
20. Not according to Max Oelschlaeger, who, in his *The Idea of Wilderness*, suggests that appreciation for the wild began in Paleolithic times. I suspect this is more of a wishful rewording of the past than it is a vindication of the Romantic intoxication with nature.
21. Immanuel Kant, *Critique of Judgment*, pp. 110–111.
22. Christopher Alexander, *The Timeless Way of Building*, p. 154.
23. Peter Wessel Zapffe, quoted in *Wisdom in the Open Air*, p. 71. This book contains the only published English translations of Zapffe's

work. An intriguing take on his ideas also appears in Herman Tennessen, "Happiness is for the Pigs."

24. Jacques Ellul, *The Technological Society*, p. 94.

25. R. L. Wing, *The I Ching Workbook*, p. 160.

CHAPTER 6: HOME AND THE WORLD

1. Immanuel Kant, *Critique of Judgment*, p. 268.

2. My interpretation differs somewhat from that of various authorities on Heidegger. In a sauna in Boulder, Colorado, Michael Zimmerman told me that releasement is not something to be accomplished with technology, but *from* technology (see his *Heidegger's Confrontation with Modernity*, p. 219). One is released toward the world through aspects of human closeness that are independent of the machine. We thus need to be freed from the shackles of Enframing technology, still using it, but not letting it rule us. (I bolted from the sweltering steam room into the icy swimming pool, not the least to release my mind into a condition conducive to thinking this over.) Though my view may deviate from Heideggerian orthodoxy, it is probably more profound to hope for the possibility of releasement *with* technology. Otherwise we would all be doomed.

On the other hand, David Michael Levin would like to simplify *Gelassenheit* into something much more palatable: "Often translated as 'releasement,' this word could be translated more helpfully as 'letting go' and 'letting be' " (*The Listening Self*, p. 227). This view is closer to the liberation found in Zen solutions that do not judge things, but take things as they are, both and neither right nor wrong, releasing the world from our valued use of it.

3. Martin Heidegger, "Memorial Address for Conradin Kreuzer," in *Discourse on Thinking*, p. 54.

4. Martin Heidegger, "Conversation on a Country Path," in *Discourse on Thinking*, p. 68.

5. Ibid., p. 69.

6. This Fragment is sometimes removed from enumerations of the words of Heraclitus, because there is "no hint of a sentential context and hence no way to construe it as meaningful." (Charles Kahn, *The Art and Thought of Heraclitus*, p. 288.) Apparently, Heidegger did not agree.

7. Ibid., p. 251.

8. Ibid., p. 252.

9. David Abram, "Time and Space: Animal Objections and Topological Reflections," pp. 10–11.

10. Paul Klee, *The Thinking Eye*, p. 10.

11. Ibid.
12. Christopher Alexander, *The Timeless Way of Building*, p. 144.
13. Ibid., p. 304.
14. Ibid., p. 548.
15. Donald Norman, *The Psychology of Everyday Things*, p. 229.
16. Ibid., p. 47.
17. Terry Winograd and Fernando Flores, *Understanding Computers and Cognition*, p. 163.
18. Ibid., p. 74.
19. Ibid., p. 165.
20. Ibid., p. 179.
21. Martin Heidegger, "The Thing," *Poetry, Language, Thought*, p. 171. Compare also Christopher Alexander's design strategy for a teapot in *Notes on the Synthesis of Form*.
22. This example is adapted from my article, "Sudden Music: Improvising Across the Electronic Abyss."
23. Emily Dickinson, *Complete Poems*, no. 501.
24. Lao Tzu, *Tao te Ching*, p. 133.
25. Ivan Illich, *Tools for Conviviality*, p. 84. ◊
26. David Ehrenfeld, *The Arrogance of Humanism*, p. 5.
27. Ibid., p. 267.

Bibliography

Abram, David. "Time and Space: Animal Objections and Topological Reflections." Unpublished, Philosophy Department, State University of New York, Stony Brook, 1988.

Adams, Robert. *Beauty in Photography: Essays in Defense of Traditional Values.* Millerton, N.Y.: Aperture, Inc., 1981.

Adas, Michael. *Machines as the Measure of Men: Science, Technology, and Ideologies of Western Dominance.* Ithaca: Cornell University Press, 1989.

Adorno, Theodor. *Aesthetic Theory.* Tr. C. Lenhardt. London: Routledge & Kegan Paul, 1984.

Alexander, Christopher. *Notes on the Synthesis of Form.* Cambridge: Harvard University Press, 1964.

————. *The Timeless Way of Building.* New York: Oxford University Press, 1979.

Allwood, Jens. "Natural Resources: Language, Beliefs, and Concepts." In *Natural Resources from a Cultural Perspective,* pp. 52–73. Stockholm: Royal Swedish Academy, 1979.

Anders, Gunther. "The Time of the End Versus the End of Time." In *A Matter of Life,* ed. Clara Urquhart, pp. 13–28. Boston: Little, Brown, 1963.

Angus, Ian H. *Technique and Enlightenment: Limits of Instrumental Reason.* Washington, D.C.: University Press of America, 1984.

Arendt, Hannah. *The Human Condition.* Chicago: University of Chicago Press, 1958.

Argüelles, José. *The Transformative Vision: Reflections on the Nature and History of Human Expression.* Berkeley: Shambhala, 1975.

Aristotle. *Introduction to Aristotle.* Ed. Richard McKeon. New York: Modern Library, 1947.

Axelos, Kostas. *Alienation, Praxis, and Techne in the Thought of Karl Marx.* Tr. Ronald Bruzina. Austin: University of Texas Press, 1976.

Bachelard, Gaston. *The Psychoanalysis of Fire.* Tr. Alan C. M. Ross. Boston: Beacon Press, 1964.

Bacon, Francis. *The New Organon.* Tr. James Spedding et al. New York: Liberal Arts Press, 1960.

———. *The Philosophical Works of Francis Bacon.* Ed. J. M. Robertson. London: George Routledge & Sons, 1905.

———. *Selected Writings of Francis Bacon.* Ed. Hugh G. Dick. New York: Modern Library, 1955.

Barnes, Jonathan. *Early Greek Philosophy.* London: Penguin Books, 1987.

Barnett, H. G. *Innovation: The Basis of Cultural Change.* New York: McGraw Hill, 1953.

Barrett, William. *The Illusion of Technique.* New York: Anchor Doubleday, 1978.

Basalla, George. *The Evolution of Technology.* New York: Cambridge University Press, 1989.

Beck, Heinrich. "Bio-social Cybernetic Determination—or Responsible Freedom?" In *Philosophy and Technology II: Information Technology and Computers in Theory and Practice,* Boston Studies in Philosophy and Science no. 90, ed. Carl Mitcham and Alois Huning, pp. 85–95. Boston: D. Reidel, 1986.

Benedikt, Michael, ed. *Cyberspace: First Steps.* Cambridge: MIT Press, 1991.

Benton, Ted. "Marxism and Natural Limits: An Ecological Critique and Reconstruction." *New Left Review* no. 178 (1989):51–86.

Berman, Marshall. *All that is Solid Melts into Air: The Experience of Modernity.* New York, Viking Penguin, 1982.

Boardman, Philip. *The Worlds of Patrick Geddes.* London: Routledge & Kegan Paul, 1978.

Bolter, J. David. *Turing's Man: Western Culture in the Computer Age.* Chapel Hill: University of North Carolina Press, 1984.

Borgmann, Albert. *Technology and the Character of Contemporary Life.* Chicago: University of Chicago Press, 1984.

Brumbaugh, Robert S. *Platonic Studies of Greek Philosophy: Form, Arts, Gadgets, and Hemlock.* Albany: SUNY Press, 1989.

Bunge, Mario. "Philosophical Inputs and Outputs of Technology." In *History and Philosophy of Technology,* ed. George Bugliarello and Dean Doner, pp. 262–281. Urbana: University of Illinois Press, 1979.

Butler, Samuel. *Erewhon* and *Erewhon Revisited.* New York: Modern Library, 1927.

———. "Darwin among the Machines." In *Notebooks of Samuel Butler,* pp. 42–47. London: A. C. Fifield, 1918.

Cage, John. *Silence.* Middletown: Wesleyan University Press, 1961.

Canetti, Elias. *Crowds and Power.* Tr. Carl Stewart. New York: Farrar, Straus, & Giroux, 1984.

Carpenter, Stanley. "A Conversation Concerning Technology: The Appropriate Technology Movement." In *Technology and Contemporary Life,* ed. Paul Durbin, pp. 87–105. Boston: D. Reidel, 1988.

Carr, Marilyn, ed. *The Appropriate Technology Reader*. London: Intermediate Technology Publications, 1985.

Channell, David. *The Vital Machine: A Study of Technology and Organic Life*. New York: Oxford University Press, 1991.

Chapius, Alfred, and Edmond Oreoz. *Automata*. Neuchatel: Editions du Griffon, 1958.

Chase, Stuart. *Men and Machines*. New York: Macmillan, 1929.

Choe, Wolhee. *Toward an Aesthetic Criticism of Technology*. New York: Peter Lang, 1989.

Cohen, Avner, and Steven Lee. "The Nuclear Predicament." In *Nuclear Weapons and the Future of Humanity: The Fundamental Question*, ed. Avner Cohen and Steven Lee, pp. 105–114. Totowa, N.J.: Rowman and Allenheld, 1986.

Colby, K. M., J. B. Watt, and J. P. Gilbert. "A Computer Method of Psychotherapy: Preliminary Communication." *Journal of Nervous and Mental Disease* 142, no. 2 (1966):148–152.

Collingwood, R. G. *The Idea of Nature*. Oxford: Clarendon Press, 1945.

Condorcet, Antoine-Nicolas de. *The Progress of the Human Mind*. Tr. June Barraclough. London: Weidenfeld and Nicolson, 1955 (1795).

Counter, S. Allen. *North Pole Legacy: Black, White, and Eskimo*. Amherst: University of Massachussetts Press, 1991.

Daumal, René. "Le non-dualisme de Spinoza ou la dynamite philosophique." In *L'evidence absurde*, Paris: Gallimard, 1972 (1934).

Derrida, Jacques. *L'Origine de la Géométrie*. Paris: Presses Universitaires de la France, 1974.

de Solla, Derek. "An Ancient Greek Computer." *Scientific American*, June 1959, pp. 60–67.

Descartes, René. *The Essential Descartes*. Ed. Margaret Wilson, tr. Elizabeth Haldane and G. R. T. Ross. New York: New American Library, 1969.

———. *Philosophical Writings*. 3 vol. Tr. John Cottingham et al. Cambridge: Cambridge University Press, 1984.

Dessauer, Friedrich. "Technology in Its Proper Sphere." In *Philosophy and Technology*, ed. Carl Mitcham and Robert Mackey, pp. 317–334. New York: Free Press, 1972.

Dewdney, A. K. *The Turing Omnibus*. Rockville, Md.: Computer Science Press, 1989.

Dewey, John. *Experience and Nature*. La Salle, Ill.: Open Court, 1929.

———. *Philosophy and Civilization*. New York: Minton, Balch, & Co., 1931.

———. *The Quest for Certainty*. London: George Allen & Unwin, 1930.

Dickinson, Emily. *Complete Poems*. Ed. Thomas Johnson. Boston: Little, Brown, 1952.

Dickson, David. *Alternative Technology and the Politics of Technical Change*. London: Fontana, 1974.

Dijksterhuis, E. J. *The Mechanization of the World Picture*. Oxford: Oxford University Press, 1961.

Drengson, Alan. "Four Philosophies of Technology." *Philosophy Today* 26, no. 2 (Summer 1982):103–117.

———. "Toward a Philosophy of Appropriate Technology." *Humboldt Journal of Social Relations* 9, no. 2 (Spring/Summer 1982):161–176.

Dreyfus, Hubert L., and Stuart E. "Making a Mind Versus Modeling the Brain." In *The Artificial Intelligence Debate: False Starts, Real Foundations*, pp. 15–44. Ed. Stephen Graubard. Cambridge: MIT Press, 1988.

Dunn, P. D. *Appropriate Technology: Technology with a Human Face.* London: Macmillan, 1978.

Ehrenfeld, David. *The Arrogance of Humanism.* New York: Oxford University Press, 1978.

Ellul, Jacques. *The Technological Society.* New York: Knopf, 1965.

———. *Introducing Jacques Ellul.* Ed. James Holloway. Grand Rapids: William Eerdmans, 1970.

Engelbart, Doug. "A Conceptual Framework for the Augmentation of Man's Intellect." In *Vistas in Information Handling*, vol. 1, ed. Howerton and Weeks, pp. 1–29. Washington: Spartan Books, 1963.

Enzensberger, Hans Magnus. *Mausoleum: Thirty-Seven Ballads from the History of Progress.* Tr. Joachim Neugroschel. New York: Urizen Press, 1976.

———. *The Consciousness Industry: On Literature, Politics, and the Media.* New York: Seabury Press, 1974.

Feenberg, Andrew. *Critical Theory of Technology.* New York: Oxford University Press, 1991.

Ferré, Frederick. *Philosophy of Technology.* Englewood Cliffs, N.J.: Prentice Hall, 1988.

Fisher, Frank. "Technology and the Loss of Self: An Environmental Concern." *Environments* 20, no. 2 (1989):1–16.

Florman, Samuel C. *The Existential Pleasures of Engineering.* New York: St. Martin's Press, 1976.

Forbes, R. J. *Man the Maker: A History of Technology and Engineering.* New York: Abelard-Schuman, 1958.

Fox, Warwick. *Toward a Transpersonal Ecology: Developing New Foundations for Environmentalism.* Boston: Shambhala, 1990.

Fraser, J. T. *Time: The Familiar Stranger.* Amherst: University of Massachusetts Press, 1987.

Gadamer, Hans-Georg. *Truth and Method.* New York: Crossroads Press, 1986.

Giedion, Siegfried. *Mechanization Takes Command: A Contribution to Anonymous History.* New York: Oxford University Press, 1948.

Gille, Bernard, et al. *The History of Techniques.* New York: Gordon & Breach Scientific Publishers, 1986.

Hall, David L. *The Uncertain Phoenix: Adventures toward a Post-Cultural Sensibility.* New York: Fordham University Press, 1982.

Hall, Edward. *The Silent Language.* New York: Doubleday, 1959.

Haraway, Donna. *Simians, Cyborgs, and Women: Reinventing Nature.* New York: Routledge, 1991.

Hardison, O. B. Jr. *Disappearing through the Skylight: Culture and Technology in the 20th Century.* New York: Viking Penguin, 1989.

———. *Entering the Maze.* New York: Oxford University Press, 1981.

Harrison, Jim. "Unimaginable Visions." *Harvard Magazine* 92, no. 2 (November 1989):21–30.

Haugeland, John. *Artificial Intelligence: The Very Idea*. Cambridge: MIT Press, 1985.

Havelock, Eric. *Preface to Plato*. Cambridge: Harvard University Press, 1963.

Heelan, Patrick. *Space Perception and the Philosophy of Science*. Berkeley: University of California Press, 1983.

Heidegger, Martin. *Being and Time*. Tr. John Macquarrie and Edward Robinson. New York: Harper & Row, 1962.

———. *Discourse on Thinking*. Tr. John Anderson and Hans Freund. New York: Harper & Row, 1966.

———. *Poetry, Language, Thought*. Tr. Albert Hofstadter. New York: Harper & Row, 1971.

———. *The Question Concerning Technology and Other Essays*. Tr. William Lovett. New York: Harper & Row, 1977.

Heim, Michael. *Electric Language: A Philosophical Study of Word Processing*. New Haven: Yale University Press, 1987.

Hero of Alexandria. *Pneumatics*. Tr. Joseph Greenwood. New York: American Elsevier, 1971.

Hickman, Larry. *John Dewey's Pragmatic Technology*. Bloomington: Indiana University Press, 1990.

———. "The Phenomenology of the Quotidian Artifact." In *Technology and Contemporary Life*, ed. Paul Durbin, pp. 161–176. Boston: D. Reidel, 1988.

Hilgartner, Stephen, et al. *Nukespeak*. New York: Penguin Books, 1982.

Hillis, W. Daniel. "Intelligence as Emergent Behavior; or, the Songs of Eden." In *The Artificial Intelligence Debate: False Starts, Real Foundations*, ed. Stephen Graubard, pp. 175–190. Cambridge: MIT Press, 1988.

Hood, Webster. "The Aristotelian vs. the Heideggerian Approach to the Problem of Technology." In *Philosophy and Technology*, ed. Carl Mitcham and Robert Mackey, pp. 347–363. New York: Free Press, 1972.

———. "Dewey and Technology: A Phenomenological Approach." *Research in Philosophy and Technology* 5 (1982):189–207.

Hughes, Thomas P. *American Genesis: A Century of Invention and Technological Enthusiasm 1870–1970*, New York: Viking Penguin, 1989.

Husserl, Edmund. *The Crisis of European Sciences and Transcendental Phenomenology*. Tr. David Carr. Evanston: Northwestern University Press, 1970.

Husserl, Intentionality and Cognitive Science. Ed. Hubert Dreyfus and Harrison Hall. Cambridge: MIT Press, 1982.

Ihde, Don. *Technology and the Lifeworld: From Garden to Earth*. Bloomington: Indiana University Press, 1990.

Illich, Ivan. "Part Moon, Part Traveling Salesman." A radio interview with David Cayley on *Ideas*. Canadian Broadcasting Corporation, Toronto, 1989.

———. *Tools for Conviviality*. New York: Harper & Row, 1973.

Jantsch, Eric. *The Self-Organizing Universe*. Oxford: Pergamon Press, 1980.

Jaspers, Karl, *The Atom Bomb and the Future of Man*. Tr. E. B. Ashton. Chicago: University of Chicago Press, 1961.

Jonas, Hans. *The Imperative of Responsibility*. Chicago: University of Chicago Press, 1984.

Kahn, Charles H. *The Art and Thought of Heraclitus: An Edition of the Frag-

ments with Translation and Commentary. Cambridge: Cambridge University Press, 1979.

Kant, Immanuel. *Critique of Judgment.* Tr. James Creed Meredith, Oxford: Clarendon Press, 1957.

Kastenholz, Hans. "From Obligation to Motivation: A Human-Ecological Approach to the Greenhouse Effect." In *Human Ecology: Strategies for the Future,* pp. 228–236. Fort Collins: Society for Human Ecology, 1991.

Klee, Paul. *The Thinking Eye.* London: Lund Humphrey, 1961.

Klein, Jacob. "On the nature of nature." *Research in Philosophy and Technology* 2 (1979):173–188.

Kotarbinski, Tadeusz. *Praxiology: An Introduction to the Sciences of Efficient Action.* Oxford: Pergamon Press, 1965.

Kroker, Arthur. *The Possessed Individual: Technology and the French Postmodern.* Montreal: New World Perspectives, 1992.

Kvaløy, Sigmund. "Complexity and Time: Breaking the Pyramid's Reign." In *Wisdom in the Open Air: The Norwegian Roots of Deep Ecology,* ed. Peter Reed and David Rothenberg, pp. 113–132. Minneapolis: University of Minnesota Press, 1992.

————. "Ambolten hvorpå selv Gudene hamrer forgjeves" [The Anvil on Which the Gods Themselves Hammer in Vain]. *Norsk filosofisk tidsskrift* 24 (1989):225–241.

La Mettrie, Julien Offray de. *Man a Machine.* Tr. Gertrude Bussey. La Salle, Ill.: Open Court, 1912 (1748).

Lang, Berel. "Genocide and Omnicide: Technology at the Limits." In *Nuclear Weapons and the Future of Humanity: The Fundamental Question,* ed. Avner Cohen and Steven Lee, pp. 115–130. Totowa, N.J.: Rowman and Allenheld, 1986.

Lanier, Jaron. "Lucy in the Sky with Computers." Interview by Adam Heilbrun. *The Boston Phoenix,* 2 March 1990, section 2.

Lao Tzu. *Tao te Ching.* Tr. R. B. Blakney. New York: New American Library, 1955.

Latour, Bruno. *Science in Action.* Cambridge: Harvard University Press, 1987.

Laurel, Brenda. *Computers as Theater.* Reading, Mass.: Addison-Wesley, 1991.

Leiss, William. *Under Technology's Thumb.* Montreal: McGill-Queens University Press, 1990.

Levin, David. *The Listening Self: Personal Growth, Social Change and the Closure of Metaphysics.* London: Routledge, 1987.

Levinson, Paul. "Information Technologies as Vehicles of Evolution." In *Philosophy and Technology II: Information Technology and Computers in Theory and Practice,* Boston Studies in Philosophy and Science no. 90, ed. Carl Mitcham and Alois Huning, pp. 29–47. Boston: D. Reidel, 1986.

————. "Intelligent Writing: The Electronic Liberation of Text," *Technology in Society* 11 (1989):387–400.

————. *Mind at Large: Knowing in the Technological Age.* Greenwich, Conn.: JAI Press, 1988.

Levy, Stephen. *Artificial Life: The Quest for a New Creation.* New York: Pantheon, 1992.

Lifton, Robert Jay. *The Future of Immortality and Other Essays for a Nuclear Age.* New York: Basic Books, 1987.

Lovejoy, A. J. "Nature as Aesthetic Norm," *Modern Language Notes* 42, no. 7 (1927):444–450.

Lyman, Francesca, et al. *The Greenhouse Trap.* Boston: Beacon Press, 1990.

McCorduck, Pamela. *Machines Who Think.* New York: W.H. Freeman, 1979.

————. *The Universal Machine: Confessions of a Technological Optimist.* New York: McGraw Hill, 1985.

MacCormac, Earl. "Men and Machines: the Computational Metaphor." *Philosophy and Technology II: Information Technology and Computers in Theory and Practice,* Boston Studies in Philosophy and Science no. 90, ed. Carl Mitcham and Alois Huning, pp. 157–170. Boston: D. Reidel, 1986.

Machiavelli, Niccolò. *The Prince.* Tr. Leo P. S. de Alvarez. Irving, Tex.: University of Dallas Press, 1980.

McCurdy, John Derrickson. *Visionary Appropriation.* New York: Philosophical Library, 1978.

McDermott, John. "Technology: Opiate of the Intellectuals." In *Technology and Man's Future,* ed. Albert Teich, pp. 130–165. New York: St. Martin's Press, 1981 (3d ed.).

————. *The Culture of Experience: Philosophical Essays in the American Grain.* New York: New York University Press, 1976.

McGinn, Robert. "What is Technology?" *Research in Philosophy and Technology* 1 (1978):179–197.

McKibben, Bill. *The End of Nature.* New York: Random House, 1989.

McLuhan, Marshall. *Understanding Media: The Extensions of Man.* New York: McGraw Hill, 1964.

McLuhan, Marshall, and Quentin Fiore. *The Medium is the Massage: An Inventory of Effects.* New York: Bantam Books, 1967.

The McLuhan Explosion, Ed. Harry Crosby and George Bond. New York: American Book Company, 1968.

Mander, Jerry. *Four Arguments for the Elimination of Television.* New York: William Morrow, 1978.

Marcuse, Herbert. *One-Dimensional Man.* Boston: Beacon Press, 1964.

Marx, Karl. *Karl Marx: Early Texts.* Tr. David McLellan. New York: Barnes & Noble, 1971.

Marx, Leo. *The Machine in the Garden.* New York: Oxford University Press, 1964.

Maslow, Abraham. *Towards a Psychology of Being.* New York: Van Nostrand, 1968.

Meeker, Joseph. "The Immanent Alliance: New Connections among Art, Science and Technology." In *Technology and Human Affairs,* ed. Larry Hickman and Azizah al-Hibri, pp. 115–122. St. Louis: C. V. Mosby, 1981.

Merleau-Ponty, Maurice. *The Phenomenology of Perception.* Tr. Colin Smith. New Jersey: Humanities Press, 1962.

Mesthene, Emmanuel. "Thinking about Technology." In *Technology and Man's Future,* ed. Albert Teich, pp. 99–129. New York: St. Martin's Press, 1981 (3d ed.).

Meyer-Abich, Klaus Michael. "What Sort of Technology Permits the Language of Nature?" *Philosophy and Technology*, Boston Studies in the Philosophy of Science no. 80, ed. Paul Durbin and Friedrich Rapp, pp. 211–232. Boston: D. Reidel, 1983.

Meyrowitz, Joshua. *No Sense of Place: The Impact of Electronic Media on Social Behavior*. New York: Oxford University Press, 1985.

Mitcham, Carl, and Robert Mackey, eds. *Bibliography of the Philosophy of Technology*. Chicago: University of Chicago Press, 1973.

———. "Philosophy and the History of Technology." In *History and Philosophy of Technology*, ed. George Bugliarello and Dean Doner, pp. 163–201. Urbana: University of Illinois Press, 1979.

Mojtabai, A. G. *Blessèd Assurance*. Boston: Houghton Mifflin, 1986.

Mokyr, Joel. *The Lever of Riches: Technological Creativity and Economic Progress*. New York: Oxford University Press, 1990.

Morris, William. *News from Nowhere*. London: Routledge & Kegan Paul, 1970 (1891).

Mumford, Lewis. "An appraisal of Lewis Mumford's *Technics and Civilization* (1934)." *Daedalus* 3 (Summer 1959):527–536.

———. *The Myth of the Machine*. New York: Harcourt, Brace and World 1967.

———. *The Pentagon of Power*. New York: Harcourt, Brace Jovanovich, 1970.

———. *Technics and Civilization*. New York: Harcourt, Brace Jovanovich, 1934.

Naess, Arne. *Ecology, Community, and Lifestyle*. Tr. and ed. David Rothenberg. New York: Cambridge University Press, 1989.

———. *Freedom, Emotion, and Self-Subsistence: The Structure of a Central Portion of Spinoza's Ethics*. Oslo: Universitetsforlaget, 1975.

Narayanan, A. *On Being a Machine*. Chichester, U.K.: Ellis Harwood, 1988.

Needham, Joseph. *Man a Machine: In Answer to a Romantical and Unscientific Treatise Written by Sig. Eugenio Rignano and Entitled "Man not a Machine."* New York: W. W. Norton, 1928.

Norman, Donald. *The Psychology of Everyday Things*. New York: Basic Books, 1988.

Nørretranders, Tor. *Det udelelige: Niels Bohrs aktualitet i fysik, mystik, og politik* [The Indivisible: Niels Bohr's Life in Physics, Mysticism, and Politics]. Copenhagen: Gyldendal, 1985.

Oelschlaeger, Max. *The Idea of Wilderness*. New Haven: Yale University Press, 1991.

Ong, Walter. *Orality and Literacy*. London: Methuen, 1982.

Oresme, Nicole. *Le Livre du ciel et du monde*. Tr. A. D. Menut. Madison: University of Wisconsin Press, 1968.

Ortega y Gasset, José. "Thoughts on Technology." In *Philosophy and Technology*, ed. Carl Mitcham and Robert Mackey, pp. 290–313. New York: Free Press, 1972.

Pacey, Arnold. *The Culture of Technology*. Cambridge: MIT Press, 1983.

———. *The Maze of Ingenuity: Ideas and Idealism in the Development of Technology*. Cambridge: MIT Press, 1976.

Papanek, Victor. *Design for the Real World: Human Ecology and Social Change*. Chicago: Academy Chicago Publishers, 1985.

Pascal, Blaise. *Pensées*. Tr. A. J. Krailsheimer. London: Penguin Books, 1966.

Patocka, Jan. "The Varna Lecture." In *Jan Patocka: Philosophy and Selected Writings*, ed. Erazim Kohák, pp. 327–339. Chicago: University of Chicago Press, 1989.

Paz, Octavio. *The Bow and the Lyre*. Tr. Ruth L. C. Simms. Austin: University of Texas Press, 1973.

Penley, Constance, and Andrew Ross, eds. *Technoculture*. Minneapolis: University of Minnesota Press, 1991.

Penrose, Roger. *The Emperor's New Mind: Concerning Computers, Minds, and the Laws of Physics*. New York: Oxford University Press, 1989.

Petroski, Henry. *To Engineer is Human: The Role of Failure in Successful Design*. New York: St. Martin's Press, 1985.

Plato. *Collected Dialogues of Plato*. Ed. Edith Hamilton and Huntington Cairns. Princeton: Princeton University Press, Bollingen Series LXXI, 1961.

Polanyi, Michael. *Personal Knowledge: Towards a Post-Critical Philosophy*. Chicago: University of Chicago Press, 1958.

Powers, Richard. *Three Farmers on Their Way to a Dance*. New York: William Morrow, 1986.

Ramelli, Agostino. *The Various and Ingenious Machines of Agostino Ramelli*. Tr. Martha T. Gnudi. Baltimore: Johns Hopkins University Press, 1976 (1588).

Rapp, Friedrich. *Analytical Philosophy of Technology*. Boston Studies in Philosophy of Science no. 63. Boston: D. Reidel, 1981.

Rheingold, Howard. *Tools for Thought: The People and Ideas behind the Next Computer Revolution*. New York: Simon & Schuster, 1985.

Rhodes, Richard. *The Making of the Atomic Bomb*. New York: Simon & Schuster, 1986.

Ricoeur, Paul. *Freedom and Nature*. Tr. Erazim Kohák. Evanston, Ill.: Northwestern University Press, 1966.

Rifkin, Jeremy. *Time Wars: The Primary Conflict in Human History*. New York: Henry Holt, 1987.

Rignano, Eugenio. *Man not a Machine: A Study of the Finalistic Aspects of Life*. London: Kegan Paul, 1926.

Rodman, John. "On the Human Question, Being the Report of the Erwhonian High Commission to Evaluate Technological Society." *Inquiry* 18, no. 2 (Summer 1975):127–166.

Romanyshyn, Robert. *Technology as Symptom and Dream*. London: Routledge, 1990.

Ronell, Avital. *The Telephone Book: Technology, Schizophrenia, Electronic Speech*. Lincoln: University of Nebraska Press, 1989.

Ross, Andrew. *Strange Weather: Culture, Science, and Technology in the Age of Limits*. London: Verso, 1991.

Rossi, Paolo. *Philosophy, Technology and the Arts in the Early Modern Era*. Tr. Salvator Attansio. New York: Harper & Row, 1970.

Rothenberg, David. "The Greenhouse from Down Deep: What Can Philosophy

Do for Ecology?" In *Human Ecology: Strategies for the Future*, pp. 243–247. Fort Collins: Society for Human Ecology, 1991.

———. "Sudden Music: Improvising Across the Electronic Abyss." *Contemporary Music Review*, October 1991.

Sarlemijn, Andries, and Peter Kroes. "Technological Analogies and Their Logical Nature." In *Technology and Contemporary Life*, ed. Paul Durbin, pp. 237–255. Boston: D. Reidel, 1988.

Scarry, Elaine. *The Body in Pain: The Making and Unmaking of the World*. New York: Oxford University Press, 1985.

Schadewaldt, Wolfgang. "The Concepts of Nature and Technique According to the Greeks." *Research in Philosophy and Technology* 2 (1979):159–171.

Schell, Jonathan. *The Fate of the Earth*. New York: Alfred A. Knopf, 1982.

Schiller, Friedrich. *On the Aesthetic Education of Man*. Tr. Reginald Snell. New York: Frederick Ungar, 1954.

Schirmacher, Wolfgang. "From the Phenomenon to the Event of Technology." In *Philosophy and Technology*, Boston Studies in the Philosophy of Science no. 80, ed. Paul Durbin and Friedrich Rapp, pp. 275–289. Boston: D. Reidel, 1983.

Schumacher, E. F. *Small is Beautiful: Economics as if People Mattered*. New York: Harper Torchbooks, 1973.

Schuurman, Egbert. *Technology and the Future: A Philosophical Challenge*. Tr. Herbert D. Morton. Toronto: Wedge Publishing Foundation, 1980.

Sheehan, James, and Morton Sosna, eds. *The Boundaries of Humanity: Humans, Animals, Machines*. Berkeley: University of California Press, 1991.

Simon, Herbert A. *The Sciences of the Artificial*. Cambridge: MIT Press, 1969.

Smith, Anthony. "Technology, Identity, and the Information Machine." *Dædalus* 115, no. 3 (Summer 1986):155–169.

Sokolowski, Robert. "Natural and Artificial Intelligence." In *The Artificial Intelligence Debate: False Starts, Real Foundations*, ed. Stephen Graubard, pp. 45–64. Cambridge: MIT Press, 1988.

Solomon-Godeau, Abigail. *Photography at the Dock*. Minneapolis: University of Minnesota Press, 1991.

Sophocles. *Antigone*. Loeb Classical Library, vol. 20. Cambridge: Harvard University Press, 1912.

Spinoza, Baruch. *Ethics*. Tr. William Hale White, ed. James Gutmann. New York: Hafner, 1949.

———. *The Correspondence of Spinoza*. Ed. A. Wolf. New York: Russell and Russell, 1966.

Stahl, Gary. "Remembering the Future." In *Nuclear Weapons and the Future of Humanity: The Fundamental Question*, ed. Avner Cohen and Steven Lee, pp. 1–37. Totowa, N.J.: Rowman and Allenheld, 1986.

Tennessen, Herman. "Happiness is for the Pigs!" *Journal of Existentialism* 7, no. 26 (Winter 1966/7):181–214.

Tichi, Cecilia. *Shifting Gears: Technology, Literature, Culture in Modernist America*. Chapel Hill: University of North Carolina Press, 1987.

Turing, Alan. "Computing Machinery and Intelligence," *Mind* 59 (1950):434–460. Reprinted in *Computers and Thought*, ed. E. A. Feigenbaum and

J. Feldman (New York: McGraw Hill, 1963) and in *The Mind's I*, ed. D. R. Hofstadter and D. C. Dennett (New York: Basic Books, 1981).
———. "Intelligent Machinery." In *Machine Intelligence*, vol. 5, ed. B. Melzer and D. Michie. Edinburgh: Edinburgh University Press, 1969.
Turkle, Sherry. "Artificial Intelligence and Psychoanalysis: A New Alliance." In *The Artificial Intelligence Debate: False Starts, Real Foundations*, ed. Stephen Graubard, pp. 241–268. Cambridge: MIT Press, 1988.
———. *The Second Self: Computers and the Human Spirit.* New York: Simon & Schuster, 1984.
Usher, Abbott Payson. *A History of Mechanical Inventions.* New York: Dover Publications, 1988 (1954).
Vartanian, Aram. *La Mettrie's L'Homme Machine: A Study in the Origins of an Idea.* Princeton: Princeton University Press, 1960.
Vives, Juan Luis. *De causis corruptarum artium.* Basileae: 1555.
von Bertalanffy, Ludwig. *General Systems Theory.* New York: George Braziller, 1968.
von Weizsäcker, Carl Friedrich. *The Unity of Nature.* New York: Farrar, Straus, & Giroux, 1980.
Weart, Spencer R. *Nuclear Fear: A History of Images.* Cambridge: Harvard University Press, 1988.
Weiner, Jonathan. *The Next One Hundred Years: Shaping the Fate of Our Living Earth.* New York: Bantam Books, 1990.
Weizenbaum, Joseph. *Computer Power and Human Reason: From Judgment to Calculation.* San Francisco: W. H. Freeman, 1976.
Wetlesen, Jon. *The Sage and the Way: Studies in Spinoza's Ethics of Freedom.* Oslo: Institut for Filosofi, 1976.
Wiener, Norbert. *Cybernetics.* Cambridge: MIT Press, 1961 (2d ed.).
Wilhelm, Richard. Tr. *I Ching or Book of Changes.* London: Routledge, 1951.
Willey, Basil. *The Seventeenth Century Background.* New York: Columbia University Press, 1933.
Wilson, Alexander. *The Culture of Nature.* Cambridge: Blackwell, 1992.
Wing, R. L. *I Ching Workbook.* Garden City: Doubleday, 1979.
Winner, Langdon. *Autonomous Technology: Technics-out-of-Control as a Theme in Political Thought.* Cambridge: MIT Press, 1977.
———. *The Whale and the Reactor: A Search for Limits in an Age of High Technology.* Chicago: University of Chicago Press, 1986.
Winograd, Terry, and Fernando Flores. *Understanding Computers and Cognition: A New Foundation for Design.* Norwood, N.J.: Ablex Publishing Co., 1986.
Wittgenstein, Ludwig. *Tractatus Logico-Philosophicus.* Tr. David Pears and Brian McGuinness. London: Routledge & Kegan Paul, 1961 (1921).
Yovel, Yermiyahu. *Spinoza and Other Heretics.* Vol. 1: *The Marrano of Reason.* Vol. 2: *Adventures of Immanence.* Princeton: Princeton University Press, 1989.
Zapffe, Peter Wessel. *Om det tragiske* [On the Tragic]. Oslo: Aventura Forlag, 1986 (1941).
———. "The Last Messiah." In *Wisdom in the Open Air: The Norwegian*

Roots of Deep Ecology, ed. Peter Reed and David Rothenberg, pp. 40–52. Minneapolis: University of Minnesota Press, 1992.

Zimmerli, Walther Chr. "Variety in Technology, Unity in Responsibility?" In *Technology and Contemporary Life,* ed. Paul Durbin, pp. 279–293. Boston: D. Reidel, 1988.

Zimmerman, Michael. *Eclipse of the Self: The Development of Heidegger's Concept of Authenticity.* Athens: Ohio University Press, 1986.

————. *Heidegger's Confrontation with Modernity: Technology, Politics, and Art.* Bloomington: Indiana University Press, 1990.

————. "Marx and Heidegger on Technological Domination of Nature." *Philosophy Today* 23, no. 2/4 (Summer 1979):99–112.

Zuboff, Shoshana. *In the Age of the Smart Machine.* New York: Basic Books, 1988.

Index

Designer: U. C. Press Staff
Compositor: Prestige Typography
Text: 10/13 Sabon
Display: Sabon
Printer: Edwards Brothers, Inc.
Binder: Edwards Brothers, Inc.